Core Concepts for the
Civil PE Exam Breadth and Transportation Depth Practice Exams and Reference Manual: 80 Morning Civil and 40 Transportation Practice Problems

Study More Efficiently

David Gruttadauria P.E.
PECoreConcepts@gmail.com

Table of contents

I.   Core Concepts Reference Guide

<u>Morning Session</u>

1. Project Planning
    A. Quantity Take Off
    B. Cost Estimating
    C. Project Schedules
    D. Activity Identification and Sequencing
2. Means and Methods
    A. Construction Loads
    B. Construction Methods
    C. Temporary Structures
3. Soil Mechanics
    A. Lateral Earth Pressure
    B. Soil Consolidation/Foundation Settlement
    C. Effective Stress
    D. Bearing Capacity
    E. Slope Stability
4. Structural
    A. Dead and Live Loads
    B. Trusses
    C. Bending
    D. Shear
    E. Axial
    F. Deflection
    G. Beams
    H. Columns
    I. Slabs
    J. Footings
    K. Retaining Walls
5. Hydraulics and Environmental
    A. Open Channel Flow and Manning's Equation
    B. Storm Water Collection and Drainage
    C. Storm Characteristics
    D. Runoff Analysis
    E. Detention/Retention Ponds
    F. Pressure Conduits
    G. Bernoulli (Conservation of Energy)
6. Transportation/Geometrics
    A. Horizontal Curves
    B. Vertical Curves

C. Traffic Volume
D. Vehicle Dynamics
7. Materials
    A. Soils Classification
    B. Soil Properties
    C. Concrete Properties
    D. Steel Properties
    E. Material Test Methods and Specification Conformance
    F. Compaction
8. Site Development
    A. Cut and Fill
    B. Construction Site Layout and Control
    C. Construction Erosion and Sediment Control
    D. Impact of Construction on Adjacent Facilities
    E. Safety

Transportation Depth

I. Traffic Engineering (Capacity Analysis and Transportation Planning)
    A. Uninterrupted flow (e.g., level of service, capacity)
    B. Street segment interrupted flow (e.g., level of service, running time, travel speed)
    C. Intersection capacity (e.g., at grade, signalized, roundabout, interchange)
    D. Traffic analysis (e.g., volume studies, peak hour factor, speed studies, modal split)
    E. Trip generation and traffic impact studies
    F. Accident analysis (e.g., conflict analysis, accident rates, collision diagrams)
    G. Nonmotorized facilities (e.g., pedestrian, bicycle)
    H. Traffic forecast
    I. Highway safety analysis (e.g., crash modification factors, Highway Safety Manual)

II. Horizontal Design
    A. Basic curve elements (e.g., middle ordinate, length, chord, radius)
    B. Sight distance considerations
    C. Superelevation (e.g., rate, transitions, method, components)
    D. Special horizontal curves (e.g., compound/reverse curves, curve widening, coordination with vertical geometry)

III. Vertical Design
    A. Vertical curve geometry
    B. Stopping and passing sight distance (e.g., crest curve, sag curve)
    C. Vertical clearance

IV. Intersection Geometry

A. Intersection sight distance

B. Interchanges (e.g., freeway merge, entrance and exit design, horizontal design, vertical design)

C. At-grade intersection layout, including roundabouts

V. Roadside and Cross-Section Design

A. Forgiving roadside concepts (e.g., clear zone, recoverable slopes, roadside obstacles)

B. Barrier design (e.g., barrier types, end treatments, crash cushions)

C. Cross-section elements (e.g., lane widths, shoulders, bike lane, sidewalks)

D. Americans with Disabilities Act (ADA) design considerations

VI. Signal Design

A. Signal timing (e.g., clearance intervals, phasing, pedestrian crossing timing, railroad preemption)

B. Signal warrants

VII. Traffic Control Design

A. Signs and pavement markings

B. Temporary traffic control

VIII. Geotechnical and Pavement

A. Sampling and testing (e.g., subgrade resilient modulus, CBR, R-Values, field tests)

B. Soil stabilization techniques, settlement and compaction, excavation, embankment, and mass balance

C. Design traffic analysis and pavement design procedures (e.g., flexible and rigid pavement)

D. Pavement evaluation and maintenance measures (e.g., skid, roughness, rehabilitation treatments)

IX. Drainage

A. Hydrology (e.g., Rational method, hydrographs, SCS/NRCS method), including runoff detention/retention/water quality mitigation measures

B. Hydraulics, including culvert and stormwater collection system design (e.g., inlet capacities, pipe flow, hydraulic energy dissipation), and open-channel flow

X. Alternatives Analysis

A. Economic analysis (e.g., present worth, lifecycle costs)

II.   Practice Problems
III.   Solutions
IV.   Answer Key

CORE CONCEPTS REFERENCE GUIDE

MORNING BREADTH

Project Planning

Quantity Take-off Methods

Quantity take-off methods are a means for estimating the cost of each aspect of a project. A project consists of many activities and materials all of which are accounted for as items of a project. For example, a project may involve the construction of a retaining wall. There are many activities and materials associated to complete this. Some include excavation, formwork, concrete for the wall, reinforcing steel etc. When contract drawings and specifications are developed, all of these items must be identified. All items also must include a quantity associated it's them to indicate the amount or extent of work for the item. These quantities must be defined by a particular unit of measure which must be appropriate for the action or material. Taking excavation as an example of an item, there must be an amount of excavation associated with it. Since excavation involves removing a volume of material, the most appropriate unit is cubic yard or cubic feet. To estimate the cost of the project, each item has a price per unit associated with it. This price is determined by previous similar work and taking into account the specifics of the particular project. Below is an example of the breakdown of some items associated with an example retaining wall project:

Item	Unit	Quantity	Unit Price	Cost of Item
Excavation	Cu. Yard	50	100	5000
Concrete (Including Formwork and labor)	Cu. Yard	25	1000	25000
Reinforcing Steel	Lb.	500	12	6000
Backfill	Cu. Yard	40	50	2000
Drainage Pipe	Linear Ft	30	5	150

Cost Estimating

Engineering Economics is used to determine the best economic course of action when weighing construction options by incorporating the life-span of alternatives and comparing costs at equivalent times. The following chart provides the most common equations;

Converts	Formula
P to F	$(1+i)^n$
F to P	$(1+i)^{-n}$
F to A	$\dfrac{i}{(1+i)^n - 1}$
P to A	$\dfrac{i(1+i)^n}{(1+i)^n - 1}$
A to F	$\dfrac{(1+i)^n - 1}{i}$
A to P	$\dfrac{(1+i)^n - 1}{i(1+i)^n}$

P = Present Value
F = Future Value
i = Interest Rate
n = Years
A = Uniform Series Value

Project Schedules

Project schedules must be set and maintained to ensure it remains on time and on budget. To determine a project schedule, all tasks must be identified and the length of time (durations) for each task must be estimated. These tasks can then be sequenced by determining what the appropriate order of tasks are. Some tasks must be completed before others can begin. These tasks are defined as predecessors. See the example chart below indicating identified tasks, durations, and predecessors:

Task	Duration (Days)	Predecessor
A	2	
B	3	A
C	2	A
D	1	B
E	2	B, C

This information can then be visualized by producing and activity diagram. First begin by drawing tasks. Start with A:

Then determine which tasks have A as a predecessor. Draw these tasks as well with arrows indicating these tasks are connected:

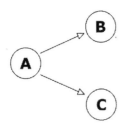

Continue in the same manor with each task. The final chart is as follows:

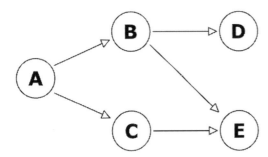

Then you can determine the critical path of the project. The critical path is defined as the sequence of tasks which would yield the shortest amount of time to complete the project. If the duration of any task on the critical path is changed, the duration of the entire project will change. In the example above, you can determine the critical path by identifying all paths and the critical one is the longest sum of duration. Therefore, the possible paths are A-B-D, A-B-E, and A-C-E which have total durations of 6, 7, and 6. Therefore the critical path is A-B-E. A change in duration of non-critical tasks will only change the project duration if the change creates a longer path than the existing critical one.

Activity Identification and Sequencing

The appropriate steps in the proper sequence need to be identified to complete a project. This involves understanding all the tasks involved in a specific project type and providing a timeline of events to properly facilitate the successful completion of the project. There are many types of projects and the specifics can vary. For the purposes of the PE Exam, it is important to have a general knowledge of common construction tasks and sequences. Below are some examples of design and construction tasks divided by when they occur in certain project phases:

Pre-Design/Design/Project Award

- Owner initiates project
- Owner hires Architect/Engineer or uses In-House Architect/Engineer
- Contract documents and specifications are developed
- Contractors bid on the project
- Project is awarded

Pre-Construction

- Contractor submittals are reviewed and approved
- Sub-Contractors hired
- Site survey, staking, and layout
- Procurement of materials

Construction

- Traffic Control, water handling, etc. installed if necessary
- Crane set up and positioning
- Temporary earth retaining systems installed if necessary
- Excavation
- Formwork or Erection
- Testing of materials
- Installation of rebar
- Pouring of concrete
- Concrete curing
- Backfill

Post-Construction

- Semi Final/Final Inspections
- Open road to traffic
- Punch-Lists
- As-Built drawings

Means and Methods

Construction Loads

Construction loads are temporary loads, occurring within the duration of a project, imposed on a structure which may be partially or fully complete. This may include materials, personnel, equipment, or temporary structures. The concern for construction loads is to understand the different types of stresses they may impose on members as opposed to the final in-place condition of those members and ensure they are designed to handle these forces.

Materials: Storage of materials is an often overlooked aspect of a project. Rebar, excavated materials, or other building materials need to be stored in an accessible location and will often impose a large additional dead load on the structure.

Temporary Structures: Temporary structures may often be needed to either provide additional support to unstable members or access for personnel to continue the erection process. Temporary structures may also be for the housing of materials or personnel.

Equipment: Equipment is often needed for various construction activities such as welding or painting procedures. The weight and distribution of these loads should be accounted for.

Cranes: Cranes can also be considered equipment however special attention should be given to the sequencing of erection based on the cranes reach.

Members in Temporary Conditions: Along with additional dead load, construction can introduce stresses into members for which they are not designed. Some examples include the erection of a precast member such as a wall panel which may be designed for compression but will see some flexure about its weak axis while it is being picked and placed. Also, the first steel girder in a bridge before it is connected to the others through diaphragms will be unstable and must be temporarily supported. In these conditions, design measures need to be taken even though they are not required for the final condition.

Construction Methods

Steel: Strong and durable material. Steel has the capabilities to be used for long span bridges and high rise buildings. Steel members are manufactured using either the hot-rolled or cold-formed methods. Steel members are provided in predetermined shapes. Some examples include W-, S-, C, and HSS-shapes. Steel is connected and constructed by the use of bolted or welded connections. The advantages of steel are again the ability to span long distances and the weight of the members compared to the strength is relatively low. Some disadvantages include the high cost, lack of ability to form unique shapes, and tendency of the material to corrode. When steel is exposed to salts, a chemical reaction occurs causing the steel to rust and even loose section properties. To counteract corrosion some preventive measures are paint systems, coating systems such as galvanizing, or weathering steel.

Reinforced Concrete: Concrete is strong in compression but weak in tension. However it has the ability to bond to reinforcing steel to appropriately resist tension. Reinforced concrete is used in buildings and shorter span bridges or certain components of bridges. Some common applications are foundation elements, bridge decks and parapets, or retaining walls. The advantage of concrete is it can be formed to any shape or aesthetic look with proper formwork and is strong in compression. The disadvantages however are that concrete has a high self-weight, will likely crack, and has a limited span length. The reinforcing in concrete can also corrode and cause pop-outs or spalls.

Precast/Prestressed/Post-Tensioned Concrete: Precast concrete is concrete which is cast somewhere other than its final location, either at a plant or another area on the construction site, and is then stripped from its forming, transported to the site, and erected. Prestressed concrete is precast concrete which has been pre-compressed using steel strands with high elasticity. The strands are tensioned to a design force before the concrete is cast. Then the concrete mix is placed and cured. The strands are then cut at the ends. Since the strands have a high elasticity, they will try to return to their original state. However, since the stands are now bonded to the concrete, there is a compressive force transferred to the concrete. This force will oppose the stresses caused by bending. Post-Tensioned concrete uses the same concept as prestressed concrete. However, the concrete member is cast first and the strands are tensioned through the member using plastic tubes embedded along the length of the member. The tube is then grouted and the strands are cut to transfer the force. Precast Concrete will have the lowest tensile capacity and therefore is used for the lowest spans. Precast is often used for compression members or bridge deck units. Prestressed Concrete is be able to span larger distances and is used for floor members. It is common in parking garages as double-tee shapes for floor members or for long span bridges with common shapes such as prestressed bulb tees. Post tensioned concrete is not as common and is used for much larger spans. Precast concrete advantages are the quality of concrete is often better under plant-controlled conditions and the construction is much quicker. The disadvantages are there is a higher cost than reinforced concrete due to shipping and erection expenses and the tendons also are susceptible to corrosion.

Wood: Relatively low strength material. Wood is often used in residential applications or for very small span bridges. Wood is extremely cheap and lightweight for erection. In addition to the low strength, wood also will deteriorate due to rot and is highly sensitive to fire damage.

Masonry: Can be reinforced or unreinforced. Masonry is also strong in compression but weak in tension. Only used in small retaining wall applications and some older bridges still are composed of masonry components.

<u>Temporary Structures</u>

Structures which are built for a specific purpose, often to facilitate an aspect of a construction project, and are removed before the conclusion of the project are temporary structures. Some examples include:

- Temporary Buildings: May be used for storage or offices during construction.
- Scaffolding: Temporary elevated platforms which provide access to perform certain tasks.
- Temporary Supports and Shoring: Often forces are introduced during construction which are not the same as the final in-place conditions. In these situations, temporary supports are needed to keep structural members stable until the construction can be complete.
- Temporary earth retaining systems: When excavation is needed, there is often not enough room to safely dig to the required depths. There may be a need to support traffic or adjacent facilities during the excavation. In these cases temporary earth retaining is required. Some examples include sheet piles, concrete blocks, or trench boxes.
- Formwork: Concrete formwork is used to place concrete to the desired shape and will remain in place until the mix has cured to the desired strength.
- Cofferdams: A wall constructed to prevent the flow of water to a specific area. Can be made of sandbags, sheet piling, or other materials.

Soil Mechanics

Lateral Earth Pressure

Rankine Active Earth Pressure

Resultant Force $R_a = \frac{1}{2} k_a \gamma H^2$ which is applied at a distance of H/3 from the base of the footing

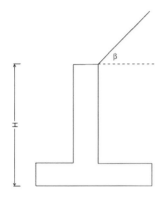

k_a = Active Earth Pressure Coefficient
γ = Density of Soil
H = Height of Retaining Wall From Base of Footing

$$k_a = \cos \beta \left(\frac{\cos \beta - \sqrt{\cos^2 \beta - \cos^2 \phi}}{\cos \beta + \sqrt{\cos^2 \beta - \cos^2 \phi}} \right)$$

β = Angle of Backfill
ϕ = Angle of Internal Friction
If the backfill is horizontal (β = 0) the equation reduces to:
$k_a = \tan^2(45° - \phi/2)$

Note: Rankine assumes the friction between the soil and wall is zero

Soil Consolidation/Foundation Settlement

Settlement is when the soil supporting the foundation consolidates which causes a decrease in volume and a drop-in elevation. This causes the foundation to no longer be fully supported and will introduce additional stress. There are three phases of settlement:

1. Immediate Settling or Elastic Settling: This settlement occurs immediately after the structure is built. The load from the structure causes instant consolidation of the soil. This is the main component of settlement in sandy soil conditions.

2. Primary Consolidation: A more gradual consolidation which is due to water leaving the voids over time. This is mostly a factor only in clayey soils.

3. Secondary Consolidation: Also occurs at a very gradual rate. This is due to the shifting and readjustment of soil grains. Most often this is the lowest magnitude of consolidation phases.

Effective Stress

The effective stress is the stress at a certain point below grade due to the weight of soil above. It is calculated as the density times the height of each level. However, If the water table is present, the density is reduced by that of water:

Effective Stress = $\Sigma H\gamma$ in dry soil and $\Sigma H(\gamma_s - \gamma_w)$ in saturated soils, where γ_w = 62.4 pcf

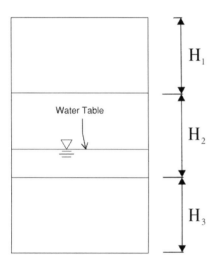

Bearing Capacity

For shallow foundations, the soil below must be suitable to support the load transferred through the footing. Different types of soils have different bearing capacities. Sand is often a good foundation material. Sand undergoes some small immediate settlement and then stabilizes since it drains quickly. Clay generally is poor in bearing capacity. Clays do not drain quickly and will retain water for longer periods of time leading to long-term settlements. Most soils in reality are some combination of sands, clays, and silts which will behave somewhere in-between sand and clay. Exceeding the allowable bearing capacity of a soil will cause shear failure or excessive settlements. Bearing capacity is determined using the Terzaghi-Meyerhof equation:

$$q_{ult} = \frac{1}{2}\gamma B N_\gamma S_\gamma + cN_cS_c + (p_q + \gamma D_f)N_q$$

q_{ult} = Ultimate Bearing Capacity
γ = Soil Density
B = Width of Footing
c = Cohesion of Soil
N_γ = Density Bearing Capacity Factor
N_c = Cohesion Bearing Capacity Factor
N_q = Surcharge Bearing Capacity Factor
p_q = Surcharge Pressure
D_f = Depth from top of Soil to Bottom of Footing
S_γ = Density Shape Factor
S_c = Cohesion Shape Factor

The following table provides bearing capacity factors based on the internal angle of friction. In between values may be interpolated:

ϕ (Degrees)	N_c	N_q	N_γ
0	5.7	1.0	0
5	7.3	1.6	0.5
10	9.6	2.7	1.2
15	12.9	4.4	2.5
20	17.7	7.4	5.0
25	25.1	12.7	9.7
30	37.2	22.5	19.7
34	52.6	36.5	35.0
35	57.8	41.4	42.4
40	95.7	81.5	100.4
45	172.3	173.3	297.5
48	258.3	287.9	780.1
50	347.5	415.1	1153.2

Shape Factors are based on the geometry of the footing where B is the width and L is the length as below:

B/L	S_c
1.0	1.25
0.5	1.12
0.2	1.05
Strip Footing	1.00
Circular	1.20

B/L	S_γ
1.0	0.85
0.5	0.90
0.2	0.95
Strip Footing	1.00
Circular	0.70

The ultimate bearing capacity then needs to be corrected for overburden to find the net bearing capacity:

$$q_{net} = q_{ult} - \gamma D_f$$

The allowable bearing capacity is then determined by dividing the net capacity by a predetermined factor of safety. A factor of safety of between 2 and 3 is common:

$$q_a = q_{net}/FS$$

Slope Stability

There are 3 types of slope failures:

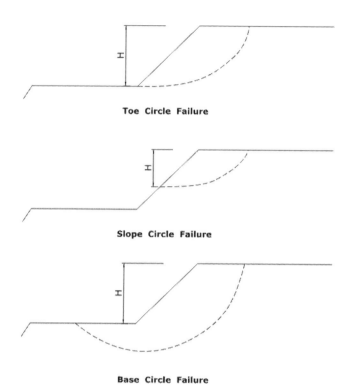

Toe Circle Failure

Slope Circle Failure

Base Circle Failure

$$F_{cohesive} = \frac{N_o c}{\gamma_{eff} H}$$

$F_{cohesive}$ = Safety factor for slope stability of cohesive soils. The minimum is often taken between 1.3-1.5
N_o = Stability number
c = cohesion (psi)
γ_{eff} = Effective soil density = $\gamma_{saturated}$ - γ_{water}
H = Depth of cut

Structural

Dead and Live Loads

Dead Loads: Loads which are permanent in the final condition of the structure. Examples include self weight and additional permanent loads (such as pavement). Dead load factors are often lower than other types of loads. This is due to the higher level of reliability being able to predict the magnitude and character of these loads.

Live Loads: Loads which will or may change over time. In general live loads represent pedestrian or vehicle loads. The load factor for live loads is often much higher due to the unpredictability.

In LRFD different types of loads are factored to represent a safety factor based on the reliability of our ability to accurately predict certain loading conditions. If only Dead and Live loads are present, the likely load combination is:

1.2D + 1.6L

Trusses

Trusses are structural members used to span long distances. Trusses are built up by members which are only in axial tension or axial compression. They can be analyzed by the method of joints as illustrated below. Consider the example truss with nodes labeled. To design, the axial force in each member must be determined. If we wanted to find the force in member BD, first like a typical beam, the reactions at A and B can be found by summing forces.

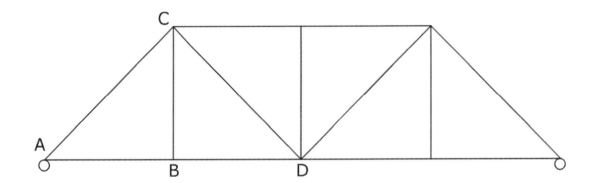

Then take a free body diagram of only the joint at A. This is shown below. In summing vertical reactions and since the reaction at A was found, the force in AC can be determined. Then there are only 2 horizontal forces and the force in AB can be found:

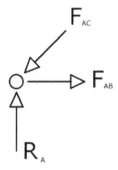

Then take a free body diagram of Joint B as shown below. Since there are only two horizontal forces, the axial force in member BD can then be determined:

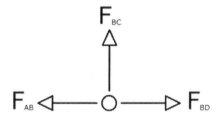

The remainder of the truss can be analyzed similarly.

Zero force members:

When determining how many 0-Force members a truss has, analyze each joint individually as a free body diagram and follow these guidelines:

1. In a joint with 2 members and no external forces or supports, both members are 0-force
2. In a joint with 2 members and external forces, If the force is parallel to one member and perpendicular to the other, then the member perpendicular to the force is a 0-force member.
3. In a joint with 3 members and no external forces, if 2 members are parallel then the other is a 0-force member

All other members are non-zero.

<u>Bending Stress</u>

Bending Stress

Mc/I

M = Applied Moment

c = Distance from the Centroid of the Cross Section to the Desired Location of Stress

I = Moment of Inertia of the Cross Section

Shear Stress

Shear stress at any point along a beam is the shear at that point over the area.

$$\tau = V/A$$

V = Shear at the point of interest

A = Cross sectional area

There is also horizontal shear stress due to bending

Horizontal shear stress $\tau = VQ/Ib$

V = Applied Shear Force (kips)

Q = First Moment of the Desired Area = ay.

a = Cross Sectional Area from Point of Desired Shear Stress to Extreme Fiber (in^2)

y = Distance from Centroid of Beam to Centroid of Area "a" (in)

I = Moment of Inertia of Beam (in^3)

b = Width of Member (in)

Axial Stress

Axial Stress:

P/A

P = Applied Force

A = Cross Sectional Area

Deflection

Deflection is the degree to which an element is displaced under load. Common equations for the maximum deflection of beams can be found in the Beam Chart

Beams

The chart below shows reactions, moments, and max deflections for common beam types:

Beam Type	Reaction	Maximum Moment	Maximum Deflection
Simply Supported w/ Single Point Load at Center	$\dfrac{P}{2}$	$\dfrac{PL}{4}$	$\dfrac{PL^3}{48EI}$
Simply Supported w/ Uniform Distributed Load	$\dfrac{wL}{2}$	$\dfrac{wL^2}{8}$	$\dfrac{5wL^4}{384EI}$
Cantilever w/ Single Point Load at Free End	P	PL	$\dfrac{PL^3}{3EI}$
Cantilever w/ Uniform Distributed Load	wL	$\dfrac{wL^2}{2}$	$\dfrac{wL^4}{8EI}$
One End Fixed, Supported at Other w/ Single Point Load at Center	$\dfrac{5P}{16}$, at Supported End $\dfrac{11P}{16}$, at Fixed End	$\dfrac{3PL}{16}$	$0.00932\dfrac{PL^3}{EI}$
One End Fixed, Supported at Other w/ Uniform Distributed Load	$\dfrac{3wL}{8}$, at Supported End $\dfrac{5wL}{8}$, at Fixed End	$\dfrac{wL^2}{8}$	$\dfrac{wL^4}{185EI}$
Beam Fixed at Both Ends, Single Point Load at Center	$\dfrac{P}{2}$	$\dfrac{PL}{8}$	$\dfrac{PL^3}{192EI}$
Beam Fixed at Both Ends, Uniform Distributed Load	$\dfrac{wL}{2}$	$\dfrac{wL^2}{12}$	$\dfrac{wL^4}{384EI}$

Shear and Moment Diagrams

Shear and moment diagrams are a graphical representation of the forces applied along the length of a beam. The following rules are used to develop shear diagrams:

-A concentrated force causes a jump In the shear diagram of equal magnitude
-A distributed load causes a line in the diagram with slope equal to the distributed load
-Forces up are positive and down is negative

This is depicted graphically below:

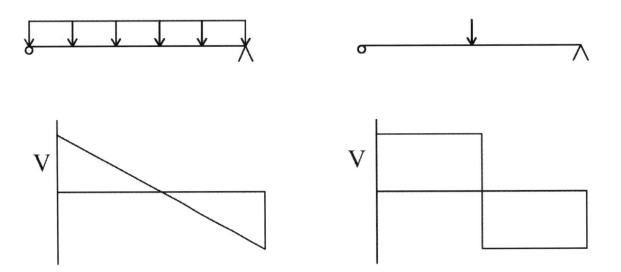

The following are rules for construction of a moment diagram:

-The moment at any point on the graph is equal to the area under the shear diagram up to this point
-An isolated moment causes a jump in the diagram of equal magnitude
-The shear at any point in the beam is equal to the slope of the same point on the moment diagram
-A distributed load will cause a parabolic moment diagram curve

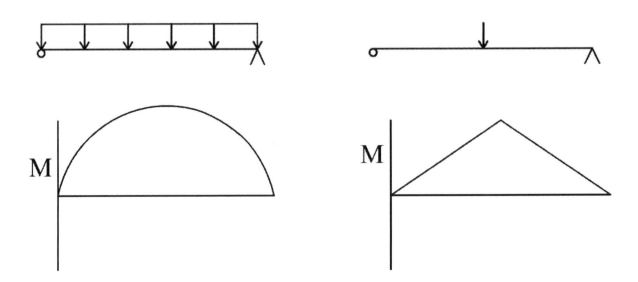

Columns

Columns are vertical members used to carry the load from spanning members to the foundation. For ideal columns where the load is concentric, the Euler formula is used to determine the theoretical maximum load:

Critical Load

$$P_{Cr} = \frac{\pi^2 EI}{(KL)^2}$$

E = Modulus of Elasticity (psi)
I = Moment of Inertia (in^4)
K = Effective Length Factor (See chart below)
L = Effective Length (in)

Critical Stress

$$F_{Cr} = \frac{\pi^2 E}{\left(\frac{KL}{r}\right)^2}$$

Effective Length Factor Chart

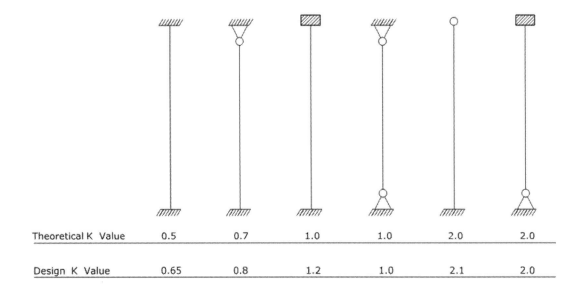

Theoretical K Value	0.5	0.7	1.0	1.0	2.0	2.0
Design K Value	0.65	0.8	1.2	1.0	2.1	2.0

Rotation Fixed and Translation Fixed

Rotation Free and Translation Fixed

Rotation Fixed and Translation Free

Rotation Free and Translation Free

Slabs

One-way slabs:

Slabs are structural elements whose length and widths are large compared to the thickness. Slabs are often used as floors or as foundation elements.

Flexure:

Slabs must be analyzed by simplified methods due to the indeterminacy of a full analysis. The most common of which is to analyze as a 1- foot wide strip and treat the span length as a beam. Transverse reinforcement is necessary to control temperature and shrinkage.

Shear:

Shear in slabs is also determined by taking one foot wide sections to analyze as a beam. However often shear will not control.

Footings

This section will focus on shear and moment for wall footings and column spread footings.

One Way Shear:

The critical section for one-way shear is at a distance d from the support.

The force applied to the footing is assumed to spread uniformly therefore the bearing pressure q_u is the force divided by the area:

$$q_u = P/(Lb_w)$$

P = Factored Load
L = Length of Footing
b_w = Width of Footing

The shear force then is the shear resulting from the bearing pressure in the area from the critical section to the free end:

$$V_u = q_u b_w \left(\frac{L}{2} - \frac{a}{2} - d \right)$$

a = Width of Column
d = Depth to Flexural Reinforcing

Moment:

The critical section for moment is at the face of the column. The applied moment resulting from the bearing pressure is:

$$M_u = q_u \frac{\left(\frac{L}{2} - \frac{a}{2}\right)^2}{2} b_w$$

Retaining Walls

Retaining walls are built to facilitate an immediate change in elevation. Some uses are to support roadways or a need for a wide, level area to be formed from a sloping existing grade. Retaining walls are designed to resist lateral loads from active earth pressure (see geotechnical section for computation of these loads) and surcharge loads which is any additional load imposed on the soil above, which when close enough will cause an additional pressure on the load due to the distribution of this load through soil. The stem of retaining walls can be analyzed as a cantilever beam extending vertically from the footing. The footing is composed of the toe which is the portion on the side of the lower elevation of soil and the heel which is portion on the side of the higher elevation of soil. Retaining walls are analyzed on a per foot width:

$$\text{Moment at base of stem} = M_{stem} = R_{ah}y_a$$

R_{ah} = Horizontal Active Earth Pressure per ft Width
y_a = Eccentricity of Horizontal Active Earth Pressure

For shear, the critical section is a distance, d, from the base of the stem where d is the distance from the main flexural reinforcement (Heel side) to the extreme compression face (Toe side):

$$V_{stem} = R_{ah}$$

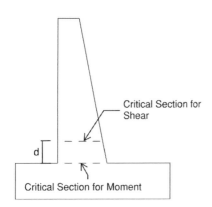

Hydraulics and Hydrology

See Transportation Depth Section

Minor Losses – Friction losses due to fittings in the line, changes in the dimensions of the pipe, or changes in direction

- Minor losses can be calculated as per the Method of Loss coefficients.
- Each change in the flow of pipe is assigned a loss coefficient, K
- Loss coefficients for fittings are most often determined and provided by the manufacturer

- Loss coefficients for sudden changes in area must be determined:

For Sudden Expansions:

$$K = \left(1 - \left(\frac{D_1}{D_2}\right)^2\right)^2$$

For Sudden Contractions:

$$K = 0.5\left(1 - \left(\frac{D_1}{D_2}\right)^2\right)$$

- $D_1 = Smaller\ diamter\ pipe$

Loss coefficients are then multiplied by the kinetic energy to determine the loss.

$$h_f = K\frac{v^2}{2g}$$

Bernoulli Equation

The Bernoulli equation for the conservation of energy states that the total energy is equal to the sum of the pressure + kinetic energy + potential energy and is conserved at any point in the system. Therefore:

$E_t = E_{pr} + E_v + E_p = p + v^2/2g + z$

E_{pr} = Pressure = p
E_v = Kinetic Energy = $v^2/2g$
v = Velocity (ft/s)
g = Acceleration Due to Gravity (32.2 ft/s^2)
E_p = Potential Energy = z = Height above point of interest to surface of water (ft)

Transportation and Geometrics

Horizontal Curves and Vertical Curves

See Transportation Depth Section

Traffic Volume

The following are traffic volume factors. It is important to note for how many lanes and directions the values represent:

Average Daily Traffic, ADT: Average number of vehicles per day over a given time period.

Average Annual Daily Traffic, AADT: Average number of vehicles per day over a year. Typically, it is calculated by dividing the total volume of vehicles in a year by 365 days.

Average Daily Truck Traffic, ADTT: Average number a trucks per day in a given time period.

Design Hour Volume, DHV: The hour of volume used in design.

K-factor, K: The ratio of the Design Hour Volume to the Average Annual Daily Traffic (DHV/AADT).

Directional Factor, D: Percentage of volume for the dominant direction of traffic during peak flow.

Directional Design Hour Volume, DDHV: The product of the Directional Factor and the Design Hour Volume.

Rate of Flow, v: Equivalent hourly rate at which vehicles pass a given point during a given time interval. The time frame is often taken as 15 min.

Design Capacity: Maximum volume a given roadway can handle.

Ideal Capacity, c: Ideal amount of volume for a given roadway. For freeways this is often taken as 2400 passenger cars per hour per lane (pcphpl).

Volume to capacity ratio: Volume over capacity, v/c

Peak Hour Factor, PHF: Ratio of the peak hour volume to the peak rate of flow in that hour:

$$PHF = \frac{V_{vph}}{4V_{15\,min,peak}}$$

28

Vehicle Dynamics

The distance it takes a driver to stop after recognizing an obstruction is the sum of two components. The first is before breaking and the second is after breaking.

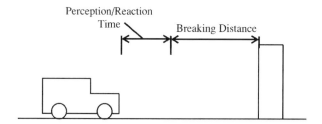

This is represented in the following equation:

$$S_{stopping} = vt_p + s_b$$

The first component assumes the velocity is constant during the perception reaction time which is the time it takes the driver to recognize the obstruction and begin the breaking process. This is calculated by:

vt_p where

v = Velocity (ft/s)
t_p = Perception Reaction Time (seconds)

The stopping distance during the breaking process is the following:

$$s_b = v^2_{mph}/(30(f+G))$$

v = Velocity (miles per hour)
f = Friction Factor with the Pavement
G = slope (positive for uphill and negative for downhill)

To convert the velocity from ft/s to mi/hour where applicable multiply by:

(3600 s/hr)/(5280 ft/mi)

Materials

Soil Classification

There are two types of common soil classification system. First is the AASHTO:

	Granular Materials (35% or less passing no. 200 sieve)								Silt-Clay materials (more than 35% passing no. 200 sieve)				
	A-1		A-3	A-2				A-4	A-5	A-6	A-7	A-8
	A-1-a	A-1-b		A-2-4	A-2-5	A-2-6	A-2-7				A-7-5 or A-7-6	
Sieve analysis: % passing:													
no. 10	50 max											
no. 40	30 max	50 max	51 min									
no. 200	15 max	25 max	10 max	35 max	35 max	35 max	35 max	36 min	36 min	36 min	36 min	
Characteristics of fraction passing no 40:													
LL: liquid Limit				40 max	41 min	40 max	41 min	40 max	41 min	40 max	41 min	
PI: Plasticity Index	6 max		NP	10 max	10 max	11 min	11 min	10 max	10 max	11 min	11 min	
Usual types of significant constituents	Stone fragments gravel and sand		Fine Sand	Silty or clayey gravel and sand				Silty Soils		Clayey Soils		Peat, highly organic soils
General subgrade rating	Excellent to good							Fair to poor				Unsatisfactory

Second is the Unified Soil Classification System (USCS):

Major Division		Group Symbol	Laboratory classification criteria		Soil Description
			% finer than 200 sieve	Supplementary requirements	
Coarse-grained (over 50% by weight coarser than no. 200 sieve)	Gravelly soils (over half of coarse fraction larger than no. 4)	GW	0-5[a]	D_{60}/D_{10} greater than 4. D^2_{30} % / $(D_{60}D_{10})$ between 1 and 3	Well graded gravels, sandy gravels
		GP	0-5[a]	Not meeting above gradation requirement for GW	Gap-graded or uniform gravels, sandy gravels
		GM	12 or more[a]	PI less than 4 or below A-Line	Silty gravels, silty-sandy gravels
		GC	12 or more[a]	PI 7 and above A-Line	Clayey gravels, clayey sandy gravels
	Sandy soils (over half of coarse fraction finer than no. 4)	SW	0-5[a]	D_{60}/D_{10} greater than 6. D^2_{30} % / $(D_{60}D_{10})$ between 1 and 3	Well-graded, gravelly sands
		SP	0-5[a]	Not meeting above gradation requirement for SW	Gap-graded or uniform sands, gravelly sands
		SM	12 or more[a]	PI less than 4 or below A-Line	Silty sand
		SC	12 or more[a]	PI over 7 and above A-Line	Clayey sands, clayey gravelly sands
Fine-grained (over 50% by weight finer than no. 200 sieve)	Low compressibility (liquid limit less than 50)	ML	Plasticity chart		Silts, very fine sands, silty or clayey fine sands, micaceous silts
		CL	Plasticity chart		Low plasticity clays, sandy or silty clays
		OL	Plasticity chart, organic odor or color		Organic silts and clays of high plasticity
	High Compressibility (liquid limit 50 or more)	MH	Plasticity chart		Micaccous silts, diatomaceous silts, volcanic ash
		CH	Plasticity chart		Highly plastic clays and snady clays
		OH	Plasticity chart, organic odor or color		Organic silts and clays of high plasticity
Soils with fibrous organic matter		Pt	Fibrous organic matter; will char, burn, or glow		Peat, sandy peats, and clayey peat

[a]For soils having 5-12% passing the no.200 sieve, use a dual symbol such as GW-GC

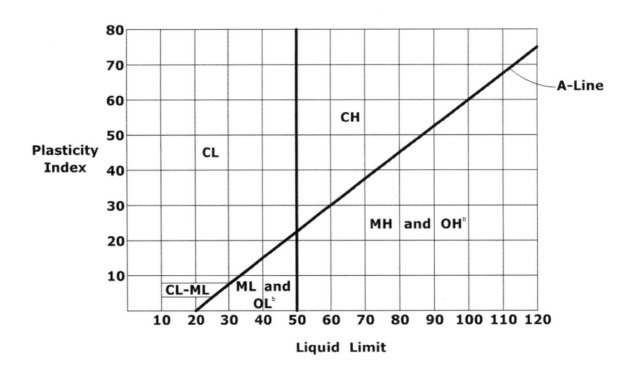

Soil Properties

The strength of soil is often determined by the standard penetration test. This measures the resistance to penetration using a standard split spoon sampler which is hit by a 140 lb hammer dropped from 30" high. The number of blows required to drive the sampler 12" after an initial penetration of 6" is referred to as the N-Value.

Permeability of a soil is a measure of continuous voids. The flow rate of water through soil depending on its permeability can be measured by Darcy's Law:

$$Q = KiA_{gross}$$

Q = Flow rate (cfs)
K = Coefficient of permeability (ft/sec)
i = Hydraulic Gradient

Phase Relationships

W_V = Weight of voids
W_w = Weight of water
W_s = Weight of solids
W_T = Total weight
V_V = Volume of voids
V_w = Volume of water
V_s = Volume of solids
V_T = Total Volume

Void ratio = $e = V_V/V_s$
Porosity = $n = V_V/V_T$
Degree of Saturation = $S = V_w/V_v$ x 100%
Moisture content = $w = W_w/W_s$
Dry Unit weight = density = γ = Weight/Volume
SG = Specific Gravity = γ_s/γ_w
γ_w = 62.4 lb/ft^3 (constant)

A saturated sample indicates the volume of the voids = volume of water
A dry sample indicates the volume of voids includes no water

Concrete Properties

Concrete consists of cement, coarse aggregate, fine aggregate, and water. Additionally, concrete may contain admixtures to enhance a certain desirable aspect of the target product. Some properties of concrete include:

Concrete Strength, f'_c: The design compressive strength of the concrete. In general this will range from 3000 to 6000 psi. However strengths can be much higher such as 20,000 psi with proper mixing and additives.

Modulus of Elasticity (Normal weight concrete), $E_c = 57,000\sqrt{f'_c}$

Modulus of Rupture or the tensile strength, $f_r = 7.5\sqrt{f'_c}$ While the tensile strength of concrete is ignored in flexure, this is often used in cracking analysis.

Water to Cement ratio, w/c, is the amount of water to the amount of cement in a given mix. In general the w/c ratio is inversely proportional to the strength since the higher amount of cement, the stronger the mix.

Cement Types:

Type I – General use cement. When special properties are not desired Type I can be used.

Type II – Used in areas where sulfate attack is a concern. This is often in areas exposed to groundwater such as drainage structures. Type II will cure at a slower rate and therefore produce less heat than other types and gain strength at a slower rate.

Type III – High early strength concrete. As opposed to type II or IV, a large amount of heat is released quickly and therefore is not suitable for mass-type pours. Type III is used in concrete where rapid strength gain is desirable such as precast concrete.

Type IV – Low heat of hydration. Gains strength slowly and generates a low amount of heat. Often used for mass-pours such as mat foundations or large retaining walls.

Structural Steel Properties

Yield Strength, F_y: Stress at which the steel will yield and begin to cause permanent deformations.

Ultimate Strength, F_u: Stress at which the steel will fracture or fail in brittle behavior.

Modulus of Elasticity, E_s = The tendency of a material to deform when subjected to forces. Also it is the ratio of stress over strain. Often in structural steel it is assumed to be taken as 29,000 ksi

Ductility: Measure of a materials ability to deform before failure. Ductility is the ratio of ultimate failure strain to yielding strain.

Toughness: The ability to withstand high stresses without fracturing.

Hardness: The ability of a material to resist surface deformation.

Material Test Methods and Specification Conformance

Concrete:

Strength tests: Most often strength is determined by loading cylinders often 6" in diameter to failure and recording the results.

Slump Test: Measure of the consistency and workability of a batch of concrete. A cone about 6" in diameter on the wide end and 12" tall is filled with concrete. The filled cone is placed on the ground and then removed to allow the concrete to naturally disperse. The remaining height and diameter of the concrete mix is measured and recorded.

Steel:

Tensile Test: Axially loading a steel member to recorded the strain in the member as the load increases. From this test the yield strength, ultimate strength, and stress strain curve can be determined. When the applied stress exceeds the yield strength, the member will undergo plastic deformation and the cross sectional area will reduce until the member fractures. This is known as necking.

Fatigue Testing: Fatigue is damage caused by repeated cycles of loading. Even though the stress in fatigue is less than the yield strength of the member, the repetition over a long period of time can cause failure. A fatigue test measures the ability of a member to resist repeated cycles of stress at a given magnitude.

Scratch Hardness Test: Also known as Mohs Test. Compares the hardness of a material to that of minerals. Minerals of known and increasing hardness are used to scratch the sample and results are observed.

Charpy V-Notch Test: Measure of a member's toughness. A member is given a 45-degree notch and a pendulum is used to hit the opposite side of the member. This is performed at different heights and magnitudes until the member fails.

Compaction

Compaction is the reduction of voids in a mass of soil. The more compacted a mass of soil is, the more stable and stronger it is to support a structure. Compaction is done by placing soil in layers called lifts and using equipment to mechanically apply weight and potentially vibration to the lifts. Some types of compaction equipment are Grid Rollers for rocky soil, sheep foot rollers for cohesive soils, or roller compactors with vibration capabilities for cohesion less soils.

When soil is compacted, the volume decreases. This is referred to as shrinkage. To calculate the compacted volume of a soil mass from its volume in its natural state use the following equation:

$$V_c = \left(\frac{100\% - \%shrinkage}{100} \right) V_b$$

V_c = Compacted Volume
%shrinkage = Percent Shrinkage
V_b = Volume of Soil in its Natural State

Conversely, when soil is removed from the ground there is an expansion of volume known as swelling:

$$V_l = \left(\frac{100\% + \%swell}{100} \right) V_C$$

V_l = Loose Volume
%swell = Percent Swell

Site Development

Excavation (Cut/Fill Estimates)

The most common method for determining the volume of excavation for cut and fill is the average end area method:

$$V = L(A_1+A_2)/2$$

L = Distance Between Area 1 and 2 (ft)
A_1 and A_2 = Respective Cross-Sectional Areas (ft^2)

Construction Site Layout

Construction sites are surveyed and markers are placed to indicate measurements and control points. These points are designated in the field by the use of stakes. These stakes can be called construction stakes, alignment stakes, offset stakes, grade stakes, or slope stakes depending on what they are meant to indicate. The accuracy of dimensions depends on the intent. Some accuracy requirements are shown below:

Type of Measurement	Level of Accuracy (ft)
Roadway Alignment, Intersections, and Paving	0.01
Buildings and Bridges	0.01
Culvert Lengths	0.1
Grade Staking	0.1
Culvert Stations	1.0
Telephone or Power Poles	1.0

Soil Erosion and Sediment Control

During construction activities involving excavation, there can be a significant amount of soil erosion leading to a dispersion of sediment. This needs to be controlled to prevent a negative impact to the surrounding areas. There are a number of options for sediment control:

Silt Fences: Fences consisting of a geotextile fabric and posts which allow the passing of runoff water but will catch the suspended sediment. They will be placed at the bottom of slopes and/or at the perimeter of the job sites at low points.

Hay Bales: Placed at the toe of slopes to help control runoff. Bales should be embedded in the ground and anchored securely with wooden posts.

Erosion Control Fabric: Geotextile fabric used for the control of erosion on steep slopes. Often these are used on piles of excavated material.

Temporary Seeding and Mulch: This involves seeding and mulching slopes to create growth that can control erosion due to the roots holding together soil. This is often used as a permanent measure for cut slopes.

Slope drain: A drain constructed to direct water to a specified area. The drain can be constructed with numerous materials such as plastic or metal pipes and concrete or asphalt. Drains must be properly anchored to resist forces from the flow of water. The outlet often is required to slow the flow of water by using energy dissipaters such as riprap.

Sediment Structure: An energy dissipating structure often made of rocks used to slow the flow of water and catch sediment.

Temporary Berm: A hill constructed of compacted soil to prevent runoff flowing in a specific direction. Berms are placed either at the top or bottom of slopes.

Impact of Construction on Adjacent Facilities

Construction can have a negative effect on surrounding properties and areas. These issues need to be anticipated and mitigated as possible. Some of the concerns include:

Construction Noise: OHSA sets maximum decibel limits on daily sound exposure. In the United States, this is typically 90 dBA for the eight-hour noise level

Runoff and Sediment: Construction sites, especially those involving excavation, can change the dynamics of runoff and drainage. See the section on Soil Erosion and Sediment Control for more details.

View: Construction projects often change the landscape of the affected area. This may have an impact on the look and feel of an area. The needs of adjacent properties may need to be considered for these changes.

Rights of Way: Often, land which is not owned by the owner of the project is necessary for the final or temporary conditions. In these cases, land needs to be acquired temporarily or permanently to complete the work. The owner of the project and of the land must come to an agreement to allow use of the property

Economic or social impact: Construction during and after may impact the access or desirability of a business or residential area. Consideration should be taken to limit the impacts to businesses or residents. For example, a bridge detour may cut off access to a restaurant which collects patrons mostly from tourists passing the effected route. The owner would then be compensated for the loss of business.

Safety

Safety is extremely important for construction sites. The *OSHA CFR 29 Part 1910 and Part 1926: Occupational Safety and Health* provides requirements for all types of construction situations and is recommended to use for the exam. Some of the highlights which you should further familiarize yourself with include:

- Excavation Safety: Except for excavations in rock, anything deeper than 5 ft must be stabilized to prevent cave-in. This may be achieved by providing appropriate earth retention systems or sloping at appropriate rates. This is determined by the depth of excavation, soil type, and other requirements.
- Fall protection: Drop-offs must be protected from fall based on the height of the drop. Some examples of protection include temporary fences, nets, or lifelines.
- Roadside Safety: Construction sites adjacent to traffic must be sufficiently protected from impact. At higher speeds concrete barriers may be needed also known as temporary precast concrete barrier curbs (TPCBC). At lower speeds it may be acceptable to provide barrels or cones to delineate the work area.
- Power line Hazards: For power lines which are electrified, all construction activities must be a minimum distance from the lines. This is based on the voltage of the lines. Typically the safe operational distance is 10 ft. for lines less than 50 kV and typically 35 ft. for lines greater than 50 kV.
- Confined Spaces: Anyone required to enter confined spaces must be appropriately trained and equipped. Oxygen must be monitored and kept at an acceptable level.
- Personal Protective Equipment (PPE): Equipment required by any personnel present on a job site. The main aspects are acceptable head protection and steel toed shoes.

TRANSPORTATION DEPTH

I. Traffic Engineering (Capacity Analysis and Transportation Planning)

A. Uninterrupted flow (e.g., level of service, capacity)

The capacity of a roadway, for a given stretch of road with defined characteristics, is a measure of the amount of traffic it can handle to maintain design speeds. The Highway Capacity Manual (HCM) is used for guidelines on the analysis of roadway capacity. Roadways must first be classified into one of two categories: Uninterrupted or interrupted flow. As the name suggests, uninterrupted flow includes roads where there is no disruption of the traffic from intersections or traffic control measures. These are typically highways or freeways. Interrupted flow is the opposite in which there are locations in which the traffic is controlled. Interrupted flow will be discussed in the next section below.

To classify roads by how they perform, the HCM has established a metric called Level of Service. A roadway can be rated from A being the best to F being the worst. The level of service is determined from charts in the HCM and is a function of the calculated density of the roadway. For freeways and multilane highway segments use HCM exhibit 12-15.

The density is determined from the following equation:

$$D = \frac{v_p}{S}$$

D = Density (pc/mi/ln)
v_p = Peak flow rate (pc/hr/ln)
S = Speed (mph)

The peak flow rate is determined from the following:

$$v_p = \frac{V}{(PHF)Nf_{HV}f_p}$$

PHF = Peak hour factor
V = Volume (vph)
N = Number of lanes
f_{HV} = Heavy vehicle factor
f_p = Driver population factor, taken between 0.85 to 1.0 with 1.0 typically if no specifics are provided

$$f_{HV} = \frac{1}{1 + P_T(E_T - 1)}$$

40

The heavy vehicle factor converts flow of trucks and buses into passenger car equivalents. The calculation of the variables is based on the characteristics of the roadway. They can be classified into either general terrain segments or individual segments. The general terrain is applicable for grades up to 3% for lengths of 0.25 to 1.0 miles. The terrain is then classified as level or rolling. E_T is then determined in exhibit 12-25.

P_T is the proportions of trucks and buses.

Free flow speed is the speed of the traffic flow when the volume is low enough to not impede the speed of the vehicles. This can be calculated off of a Base Free Flow Speed. This is the speed of a roadway under perfect geometric conditions:

For Freeways HCM Eq. 12-2:

$$FFS = BFFS - f_{LW} - f_{RLC} - 3.22TRD^{0.84}$$

For Highways HCM Eq. 12-3:

$$FFS = BFFS - f_{LW} - f_{TLC} - f_M - f_A$$

f_{LW} = Lane width adjustment factor, exhibit $12 - 20$
f_{RLC} = Right Side Lateral clearance adjustment factor, exhibit $12 - 21$
f_{TLC} = Total Lateral clearance adjustment factor, exhibit $12 - 22$
f_M = Median adjustment factor exhibit $12 - 23$
f_A = Access point density factor exhibit $12 - 24$
TRD = Total Ramp Density exhibit $(\frac{ramps}{mile})12 - 32$

B. Street segment interrupted flow (e.g., level of service, running time, travel speed)

Interrupted flow conversely to uninterrupted flow contains some restriction for the analysis of the capacity of a segment of roadway. This includes intersections both signalized and unsignalized, roundabouts, urban street flow, and pedestrians.

Because of these restrictions the traffic can often not reach the free flow speed and instead can only reach a running speed. This is the speed at which a vehicle is able to travel when accounting for the factors created by the interruption of flow. This speed can be calculated from the following equation HCM 18-48:

$$S_{RT} = min\left(S_R \ or \ \frac{61}{1 + e^{-1} + \frac{1185N_{ts}}{L}} \right)$$

L = Segment length (ft)
N_{ts} = Number of stops within segment
$$S_R = \text{Vehicle running speed} = \frac{3600L}{5280t_R}$$
The running speed can then be used to determine the running time HCM eq 18-53:

$$t_{RT} = \frac{3600L}{5280S_{RT}} + \sum_1^{N_{ts}} d_{ts}$$

d_{ts} = Sum of delays at each stop along the segment

The travel speed is then the distance traveled divided by the running time:

$$S_T = \frac{3600L}{5280(t_{RT} + d_t)}$$

$$d_t = \text{Rough vehicle delay} = t_l 60 \frac{L}{5280}$$
t_1 = Transit vehicle running time loss Exhibit 18-27

C. Intersection capacity (e.g., at grade, signalized, roundabout, interchange)

Intersection capacity is a function of the number and type of vehicles entering each approach and the delay for each vehicle at each approach. First, calculate the saturation flow rate HCM Eq. 19-8:

$$s = s_o f_w f_{HVg} f_p f_{bb} f_a f_{LU} f_{LT} f_{RT} f_{Lfb} f_{Rfb} f_{wz} f_{ms} f_{sp}$$

s_o = Base saturation flow rate
f_w = Lane width adjustment factor HCM exhibit 19-20
f_{HVg} = Heavy vehicle adjustment factor

Downgrade approach $f_{HVg} = \dfrac{100 - 0.79P_{HV} - 2.07P_g}{100}$

Zero or upgrade approach $f_{HVg} = \dfrac{100 - 0.78P_{HV} - 2.07P_g^2}{100}$

$f_p = Parking\ factor = \dfrac{N - 0.1 - \frac{18N_m}{3600}}{N} > 0.05$

$f_{bb} = Bus\ blockage\ factor = \dfrac{N - \frac{14.4N_b}{3600}}{N} \geq 0.05$

f_a = Intersection type factor = 0.9 for urban, 1.0 otherwise
f_{LU} = Lane utilization factor, HCM factor $19 - 15$
f_{LT} = Left turn adjustment factor = $1/(1.05P_{LT})$
f_{RT} = Right turn adjustment factor = $1/E_{RT}$
f_{Rfb} = Right turn pedestrian or bicycle factor = $1.0 - P_{RT}(1 - A_{pbT})(1 - P_{RTA})$
f_{Lfb} = Left turn pedestrian or bicycle factor = $1.0 - P_{LT}(1 - A_{pbT})(1 - P_{LTA})$
f_{wz}= Work Zone adjustment factor, 1.0 if not present
f_{ms}= Downstream lane blockage factor, 1.0 if not present
f_{sp} = Sustained spillback, 1.0 if not present

See signal design section for equations related to Phase and Capacity

The Level of service can then be determined using the control delay and exhibit 19-8. The control delay is the sum of the flow rate multiplied by the delay for each approach divided by the sum of the flow rates. The equation is the following (HCM Eq. 19-29):

$$d_I = \frac{\sum d_i v_i}{\sum v_i}$$

43

Roundabouts are often a good alternative to signalized intersections. The capacity per lane adjusted for heavy vehicles is a function of the conflicting flow rate:

$$c_e = 1130e^{(-0.001)v_c}$$

c_e = Lane capacity (pcph)
v_e = Conflicting flow rate (pc)

The roundabout entry capacity should not exceed 1800 vphpl.

D. Traffic analysis (e.g., volume studies, peak hour factor, speed studies, modal split)

Volume studies as the name indicates is an analysis of a roadway or intersection by field measurements. The study most often consists of observers on site counting traffic volumes and recording the numbers. The parameters of the study need to be determined by engineering judgement based on the intention of the study. The range and duration of the study can vary to achieve these intentions. The results of the study can be used to calculate parameters used in analyses such as average daily traffic, intersection volumes, and observed speeds. Many of the values which can be observed during volume studies are covered in the morning section.

For Peak Hour Factor see the morning session

It is important to set the limits of the segment in which the speed is to be recorded. This segment needs to be determined by judgement based on the intention of the study. The average speed over a given segment can be calculated by the following equation:

$$S_{avg} = \frac{N_t L}{\sum t}$$

S_{avg} = Average speed for the given segment
L = Length of segment
N_t = Number of cars observed
t = Observed time of each vehicle

Modal split is the measure of the percentages of different modes of transportation for an observed stretch. The modes often include cars, buses, trucks, bikes and pedestrians and any other uncommon mode. It is a good representation of the distribution of traffic for a given location or stretch of roadway.

E. Trip generation and traffic impact studies

Trips are the act of a type of modal transportation leaving an origin and arriving at a destination. It is important to characterize the amount and type of trips which occur in a given area. This is a trip generation analysis. This is often used to characterize trips and observations are classified as data points for a given type of trip. These data points are then charted and fit to an equation to help approximate anticipated trips. The best fit equation can be linear or nonlinear:

$$T = y + bx \ (linear)$$

$$\ln T = y + b \ln X \ (nonlinear)$$

T = Number of trips
y = y-intercept
b = Slope of best fit line
X = Trip generation parameter

F. Accident analysis (e.g., conflict analysis, accident rates, collision diagrams)

When traffic movements create a potential for a crash, these can be reviewed as a part of a conflict analysis. This identifies all of the potential movements for an intersection or roadway and determines where there is the possibility for a crash. Conflict diagrams show all movements and the types of conflicts associated with other movements. This can be used to see where there are troublesome areas and the potential for improvements to avoid undesirable conflicts.

Accident analysis is, as the name suggests, an evaluation of the number of crashes for a given intersection or segment. This information can be used to evaluate if improvements are required. The accident rate is a ratio of the number of crashes to the exposure, which is the number of vehicles for a defined time or length of roadway:

$$R = \frac{N_{CR}}{Exposure}$$

For intersections:

$$N_{Intersection} = \frac{N_{CR} 10^6}{(AADT)N_{yr} \left(365 \frac{days}{yr}\right)}$$

For roadway segment:

$$N_{Segment} = \frac{N_{CR}10^8}{(AADT)N_{yr}\left(365\frac{days}{yr}\right)L_{mi}}$$

G. Nonmotorized facilities (e.g., pedestrian, bicycle)

The analysis of pedestrians is important to the flow of vehicle traffic, to ensure the area can handle the number of pedestrians, and to ensure safety. Just as with vehicles we can calculate the pedestrian flow rate at a given location:

$$v_P = S_P D_P$$

S_P = Walking speed
D_P = Pedestrian density

It is important to note the speed of pedestrians will decrease as the density is increased. This is because people have more trouble maneuvering and walking at a normal pace if they are obstructed by other people. The Highway Capacity Manual has a number of graphs which show the relationship between density, speed, space, and flow in exhibits 4-14/15/16/17.

A walkway or sidewalk has a certain width. However, there are often objects in the walkway which will reduce the effective width. The reduction in the original walkway width is the sum of the shy distances. Typical values for reductions can be found in HCM exhibit 24-9.

Just as in the analysis of vehicular traffic, the performance for pedestrian flow at a particular walkway or intersection can be classified by a Level of Service. The flow of pedestrians can be uninterrupted or interrupted. The unit flow rate is the determining variable for the LOS but is most often taken at 15 min intervals. The 15 min pedestrian flow rate is (HCM Eq. 24-3):

$$v_P = \frac{v_{15}}{15W_E}$$

The LOS for average walkways can be determined from HCM exhibit 24-1.

There is also platoon level of service. This accounts for the fact that pedestrians will often travel in groups. Platoon LOS can be determined from HCM exhibit 24-2.

H. Traffic forecast

Predicting traffic is important for allocating funds and prioritizing projects for the future. Often traffic can be estimated using historical data to obtain a growth rate. Future traffic can be predicted using the following equation:

$$V_{Future} = AADT(1 + P)^n$$

P = Growth rate (decimal)
n = Number of years

I. Highway safety analysis (e.g., crash modification factors, Highway Safety Manual)

The AASHTO Highway Safety Manual (HSM) provides guidelines for the prediction of crashes for a given segment or location. The frequency of crashes can be predicted by using equations called Safety Performance Functions (SPF) based on the characteristics of the roadway and the desired time period. The equations must be determined through statistical modeling and are most often based on annual traffic volume and segment length but may also include other roadway characteristics. These SPF's are used to determine a predicted crash frequency which can then be adjusted to determine the actual predicted frequency from the following equation:

$$N_{Predicted} = N_{SPF}C(CMF)$$

C = Calibration factor
CMF = Product of all Crash Modification Factors

The Crash Modification Factors (CMF) are based on proposed modifications to a site. It is the ratio of the expected crash frequency of the changed site to the crash frequency of the original condition:

$$CMF = \frac{Modified\ Crash\ Frequency}{Original\ Crash\ Frequency}$$

II. Horizontal Design

A. Basic curve elements (e.g., middle ordinate, length, chord, radius)

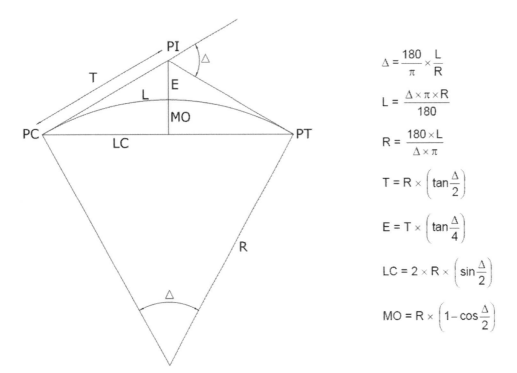

$$\Delta = \frac{180}{\pi} \times \frac{L}{R}$$

$$L = \frac{\Delta \times \pi \times R}{180}$$

$$R = \frac{180 \times L}{\Delta \times \pi}$$

$$T = R \times \left(\tan\frac{\Delta}{2} \right)$$

$$E = T \times \left(\tan\frac{\Delta}{4} \right)$$

$$LC = 2 \times R \times \left(\sin\frac{\Delta}{2} \right)$$

$$MO = R \times \left(1 - \cos\frac{\Delta}{2} \right)$$

B. Sight distance considerations

There are three types of sight distance for consideration:

Stopping sight distance is the distance it takes a driver to react to the need to stop and to apply the brakes. This is covered in detail in the morning session. AASHTO GDHS however does provide table 3-1 for quick reference based on standard values for perception reaction time and eye height.

Decision sight distance is, as it sounds, a distance required for the driver to make a decision to avoid an obstacle which has been recognized. This may involve changing lanes or going into the recoverable slope. These variables must be identified and can be classified by the GDHS into maneuver categories A through E. Table 3-3 provides decision sight distances based on the design speed and avoidance maneuver. The table values are from the following equations:

$$\textit{For A and B: } d_{ft} = 1.47 v_{mph} t_{p,sec} + 1.075 \left(\frac{v_{mph}^2}{a_{ft/s^2}} \right)$$

$$For\ C\ through\ E: 1.47v_{ft/s}t_{t,sec}$$

The last type is passing sight distance. This is the amount of distance for a vehicle to speed past a slower moving vehicle. GDHS provides Table 3-4 to determine the required distance based on design, slower, and faster vehicle speeds.

Objects or buildings which fall within a horizontal curve must be analyzed to ensure the sightline of the driver to a point further along the curve is not obstructed. This ensures appropriate horizontal clearance. The horizontal offset of a curve to an object is given by the following equation:

$$H = R\left(1 - \cos\frac{28.65S}{R}\right)$$

H = Offset
S = Horizontal sight distance
R = Radius of the curve

C. Superelevation (e.g., rate, transitions, method, components)

When a vehicle travels along a horizontal curve, there is a centrifugal force acting radially out on the vehicle. Roadways along a straight alignment are sloped away from the centerline. If the cross section remains the same along a curve, the combination of the radial force and the downslope may cause cars to tip or slide off the road. To counteract this, the cross-section transitions on the outside of the curve from having a downslope to having an upslope. This sloping is called superelevation. The rate of slope necessary to safely allow the car to cross the horizontal curve is the following equation:

$$e_{decimal} = \frac{v^2_{mph}}{15R} + f$$

e = Rate of superelevation
v = Velocity
R = Radius of curve
f = Friction factor

The transition to a fully superelevated section consists of two parts. The crown runoff (tangent runout) is the length over which the roadway transitions from its normal cross section to a flat grade. The second is the superelevation runoff which is the length over which the roadway transitions from flat to super elevated. The length of the superelevation runoff can be determined from the following equation:

$$L = \left(\frac{W N_L e}{\Delta_\%}\right) b_w$$

W = Width of one lane
N_L = Number of lanes
e = Rate of superelevation
Δ = Maximum relative gradient (GDHS Table 3-15)
b_w = Lane adjustment factor (GDHS Table 3-16)

D. Special horizontal curves (e.g., compound/reverse curves, curve widening, coordination with vertical geometry)

Compound curves combine two horizontal curves together. To analyze, understand that the end of the first curve is the beginning of the second. Therefore:

$PT_1 = PC_2 = PCC$

The point in common is also called the Point of Continuing Curve (PCC)

III. Vertical Design

A. Vertical curve geometry

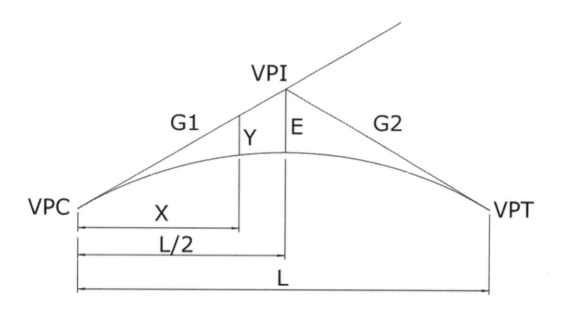

A = Gradient of the curve = $g_2 - g_1$

Elevation at a distance x from VPC = $E_{VPC} + g_1x + \frac{1}{2}(rx^2)$

$$E = \frac{AL}{800}$$

$$Y = \frac{Ax^2}{200L}$$

B. Stopping and passing sight distance (e.g., crest curve, sag curve)

For vertical curves, the slopes will have an impact on the cars ability to recognize an object and stop or properly pass. The length of the curve must be adequate for these conditions. There are two methods of determining this minimum length and their use is dependent on the variables provided.

The first method is the use of the K-value. The factor K is the ratio of the length of curve to the absolute difference in grades:

$$K = \frac{L}{|G_2 - G_1|}$$

K can also be determined from the design speed by the charts in the AASHTO Policy on the Geometric Design of Highways and Streets Table 3-34/35/36.

The AASHTO GDHS also provides equations correlating the stopping sight distance to the curve length and the difference in grade. The appropriate equation can be used depending on the curve type, stopping or passing distance, and if the stopping distance is greater than the curve length.

C. Vertical clearance

Vertical clearance is the height from the roadway to an obstruction, often a bridge above. There are two concerns when analyzing a vertical curve for clearance issues. The first is to ensure the object above the road does not inhibit the necessary sight distance. The second is to ensure the height of vehicle can safely pass under the obstruction. The required length of curve for a required clearance can be determined from the following equations:

$$S > L; L = 2S - \frac{800\left(C - \left(\frac{h_1 + h_2}{2}\right)\right)}{A}$$

$$S < L; L = \frac{AS^2}{800\left(C - \left(\frac{h_1 + h_2}{2}\right)\right)}$$

S = Sight distance (ft)
C = Clearance (ft)
h_1 = height to driver eye
h_2 = Height to object
A = Algebraic difference in grade

IV. Intersection Geometry

A. Intersection sight distance

When a vehicle is approaching or is stopped at an intersection, they must have an adequate line of sight along the perpendicular roadway to be able to safely stop or maneuver if necessary. This sight distance can be approximated by sight triangles where the hypotenuse is the required sight distance and the base is the required stopping distance. The diagram below exhibits this where X is the stopping distance of the vehicle on the major road and H is the sight distance:

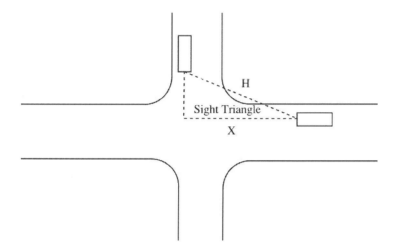

B. Interchanges (e.g., freeway merge, entrance and exit design, horizontal design, vertical design)

An interchange is a grade-separated crossing of 2 or more roadways in which ramps are used in such a manner so that the flow of traffic is not interrupted. On ramps and off ramps need to be designed such that there is enough length for acceleration and sight distance for the seamless merging of traffic. Mostly the design lengths can be determined from the appropriate tables in the GDHS.

There are a number of types of interchanges which have advantages and disadvantages based on the site constraints. Some examples include trumpet, diamond, partial and full cloverleaf, or fully directional

As with traffic signals, GDHS provides warrants for the consideration of the use of interchanges. These include:

1. Design Designation
2. Bottleneck or Spot Congestion Relief
3. Safety Improvements
4. Topography
5. User Benefits

6. Traffic Volume

C. At-grade intersection layout, including roundabouts

Intersections must be detailed to minimize disruption of traffic and to ensure a safe driving condition. To achieve this, the layout must facilitate both proper sight distances and maneuverability. Acute angles at intersections provide difficulties for both of these aspects and should be avoided as much as possible. The AASHTO Policy on the Geometric Design of Highways and Streets (GDHS) provides a wide range of tables and figures. Chapter 2 focuses on vehicle dimensions and the ability to make turns. Chapter 9 provides guidance on the geometry of the traveled way and intersections to account for minimum turning requirements.

V. Roadside and Cross-Section Design

A. Forgiving roadside concepts (e.g., clear zone, recoverable slopes, roadside obstacles)

Drivers, for a number of reasons, may veer off the road whether it be distraction, fatigue, or to avoid collision. For proper roadway design, there needs to be a minimum horizontal distance so that the driver can safely return to the roadway unharmed. This horizontal distance which begins at the edge of the roadway is called the clear distance. The AASHTO Roadside Design Guide (RSDG) provides guidelines on the safety of cars which have traveled off of the roadway.

The land just outside of the roadway may not always be flat. The slope of the clear distance has an effect on the cars ability to safely recover. Slopes less than 1 Vertical to 4 Horizontal are considered recoverable slopes since the car's ability to stop or maneuver will not be greatly affected by the slope. A non-recoverable slope is one which is steeper than 1:4. If a non-recoverable slope is present, the bottom of the slope must have a vehicle runout area which will allow the vehicle to stop. Table 3-1 of the AASHTO RSDG can be used to determine minimum clear distances based on slopes and design speeds.

When traveling on a horizontal curve, the cars traveling along the outside of the curve will struggle to recover more-so than a straight roadway due to the centrifugal force. Therefore, an adjustment factor needs to be applied to the clear zone on the outside of the curve only. The adjustment factor is found in table 3-2 of the AASHTO RSDG.

B. Barrier design (e.g., barrier types, end treatments, crash cushions)

Often objects outside of the roadway must fall within the clear zone. A barrier must be provided to both protect the object and prevent the vehicle from a collision. An appropriate barrier will minimize the damage to the vehicle and safely redirect it onto traffic. The runout length, L_R, is the minimum distance away from an object that a vehicle may leave the roadway and strike the object. This will define the length of barrier needed. AASHTO RSDG Table 5-10b provides minimum values based on volume and design speeds. Barriers which are too close to the roadway may be troublesome to drivers and cause them to slow down. To prevent this, a minimum shy distance is provided in RSDG Table 5-7. The geometry of a barrier must be determined for a safe condition by the following equations:

Barrier with flare:

$$X = \frac{L_A + \frac{b}{a}L_1 - L_2}{\frac{b}{a} + \frac{L_A}{L_R}}$$

Barrier without:

$$X = \frac{L_A - L_2}{\frac{L_A}{L_R}}$$

Lateral offset:

$$Y = L_A - \frac{L_A}{L_R}X$$

L_A = Distance from edge of road to back edge of object
b = Rise of taper slope
a = Run of taper slope
L_1 = Length from object to beginning of flare
L_2 = Distance from edge of road to face of barrier
L_R = Runout Length

Crash attenuators can be used to prevent vehicles from crashing directly into an object or from entering an area which would be unsafe for the driver or pedestrians. When the vehicle strikes the attenuator, it begins to decelerate at a rate of the following equation:

$$d = \frac{v^2}{2Lx}$$

d = Deceleration rate (ft/s^2)
v = Velocity (ft/s)
L = Length of attenuator (ft)
x = Attenuation efficiency factor

The stopping force then is:

$$F = \frac{SF(w)d}{g}$$

55

F = Stopping force (lbs)
w = weight of vehicle (lbs)
d = Deceleration rate
g = Force due to gravity (32.2 ft/s^2)
SF = Safety factor

C. Cross-section elements (e.g., lane widths, shoulders, bike lane, sidewalks)

While a roadway often has to fit the area and purpose of its proposed location, the geometric features must meet certain minimum and maximum values. The Policy on Geometric Design of Highways and Streets provides a large number of requirements for the design of a roadway or walkway cross section. For the PE exam it is best to become familiar with the location of these requirements and most importantly be able to find them quickly since it is unreasonable to be expected to memorize all values.

D. Americans with Disabilities Act (ADA) design considerations

The American Disabilities Act of 1990 outlines the requirements for structures to ensure proper treatment of individuals with disabilities. The guidelines outline many topics including parking, ramps, egress and others and the requirements which must be met to ensure the proper accessibility and safety. For the PE exam you will likely be asked a question or two requiring you to lookup certain aspects of the code. You should not spend excessive amounts of time reading the code but be familiar with the sections and be able to navigate and find information quickly.

VI. Signal Design

A. Signal timing (e.g., clearance intervals, phasing, pedestrian crossing timing, railroad preemption)

Traffic signals must be designed to allow enough time for each traffic movement at the intersection to handle the volume in that direction. Each individual set of traffic movements are grouped together by movements which can safely coexist. The signal then changes and another group of movements is allowed to proceed. The time for each group is a phase. The time it takes to go through all phases, is the cycle. The phase length is the following equation (HCM Eq. 19-2):

$$D_p = l_1 + g_s + g_e + Y + R_c$$

l_1 = Time for drivers to react to green
g_s = Time required to clear the queue
g_e = Time extension for random vehicles
Y = Yellow light time
R_c = Red light clearance time

The effective green or the time in which vehicles are actually flowing can be taken as (HCM Eq. 19-3):

$$g = g_s + g_e + e$$

e = Effective green extension (often taken as 2 seconds)

The capacity of pedestrians entering a signalized intersection must be determined to ensure there is sufficient time to handle the volume of people. The number of pedestrians which can cross at a single cross walk during a single cycle is determined from the following (HCM Eq. 19-66):

$$N_p = \frac{N_{do}(C - g)}{C}$$

C = Cycle length (s)
g = Green time (s)
N_{do} = Pedestrian per cycle (People/cycle)

The time to have a cross walk cleared can be determined from (HCM Eq. 19-64):

$$t = 3.2 + \frac{L}{S_p} + 2.7\frac{N_p}{W}$$

L = Length of crossing (ft)
S_p = Pedestrian design speed (ft/s)
W = Width of cross walk (ft)

B. Signal warrants

Intersections which do not include traffic signals may be analyzed to determine if one is required. The Manual on Uniform Traffic Control Devices (MUTCD) provides eight different scenarios which can be used to identify intersections which may require a signal to be installed. The following are the eight warrants:

- Warrant 1, Eight-Hour Vehicular Volume
- Warrant 2, Four-Hour Vehicular Volume
- Warrant 3, Peak Hour
- Warrant 4, Pedestrian Volume
- Warrant 5, School Crossing
- Warrant 6, Coordinated Signal System
- Warrant 7, Crash Experience
- Warrant 8, Roadway Network

- Warrant 9, Intersection Near a Grade Crossing

VII. Traffic Control Design

A. Signs and pavement marking

MUTCD Chapter 3 provides requirements for signs and pavement markings. As with cross sectional elements, become familiar with this chapter and how to locate requirements quickly.

B. Temporary traffic control

When work in the roadway is necessary, traffic must be shifted and directed away from the work zone in a safe manner for both the flow of traffic and the workers in the zone of construction. Tapered traffic control devices are used to direct traffic away from the work zone. MUTCD provides equations for the suggested taper length.

$$L = \frac{WS^2}{60}; S < 40 \ mph$$

$$L = WS; S > 40 \ mph$$

W = Width of work zone
S = Design speed (mph)

However, the length L must be adjusted depending on the type of taper as per MUTCD Table 6C-3.

VIII. Geotechnical and Pavement

A. Sampling and testing (e.g., subgrade resilient modulus, CBR, R-Values, field tests)

First, we will discuss the layers of soil beneath the pavement. Below the top layer of either a wearing surface or rigid pavement is a layer called subbase. Subbase often consists of crushed stone material. The load is transferred through the subbase to the subgrade. Which is the soil native to the area. To effectively design the pavement, we must understand the characteristics of the subgrade.

The soil resilient modulus is essentially the modulus of elasticity of the soil. It can be determined either through lab tests, using samples, or estimated through soil strength parameters. There are two types of soil strength parameters, the California Bearing Ratio (CBR) and the Soil Resistance Factor (R). These can be correlated to the resilient modulus by the following equations from the AASHTO GDPS

$$M_R = 1500(CBR)$$

58

$$M_R = 1000 + 555R$$

B. Soil stabilization techniques, settlement and compaction, excavation, embankment, and mass balance

See morning session for soil types and handling

During the construction or rehabilitation of roadway projects, there is often a significant amount of grade change that may result in cut or fill volumes. Mass Diagrams are a graphical representation of the excavation or fill along a baseline. The cumulative volume in either cut or fill is shown on the Y-axis as a function of the length along the baseline. The mass diagram is often paired with or proceeded by a profile diagram. This is an elevation graph along the length of the baseline showing the existing and proposed profile. It is often useful to identify the points along a mass diagram in which the cumulative volume is zero. This represents a point where the cut and fill are equal cumulatively to that point. These points are called balance points.

C. Design traffic analysis and pavement design procedures (e.g., flexible and rigid pavement)

The driving surface for vehicles must be durable so that it can handle the constant daily load from vehicle traffic. There are two types of pavement: flexible and rigid.

Flexible pavement is as the name suggests, a more elastic material to handle deformations due to loading and temperature changes. Bituminous pavement, or more commonly named asphalt, is the most common type of flexible pavement. It is mixed with aggregate similar to that of concrete with a bituminous binder. Bituminous can flex more under loading conditions making it more ideal in areas of less desirable subgrades. To design flexible pavement, the AASHTO Guide for the Design of Pavement Structures (GDPS) has the following layer thickness equation to determine the Structural Number. This is a number which encompasses all the properties and thicknesses of the pavement and subbase layers:

$$SN = D_1a_1 + D_2a_2m_2 + D_3a_3m_3$$

D = Layer thickness
a = Layer or strength coefficients
m = Drainage coefficients

The design of flexible pavement can also be determined from the AASHTO GDPS Figure 3.1

Rigid pavement is most often concrete. This type does not have the elasticity of the flexible pavement but is often provides more strength and durability. The AASHTO GDPS also has requirements for the design of rigid pavement. The equation for the design thickness is overly complicated for the exam and is often computed using computer models. AASHTO does however provide Figure 3.7, which has a flow chart for the design slab thickness based on a number of variables.

D. Pavement evaluation and maintenance measures (e.g., skid, roughness, rehabilitation treatments)

Skid resistance is the ability of pavement to resist sliding of the tires for a vehicle. Skid resistance can often be quantified by using a Skid Number from the following equation:

$$SN = 100\left(\frac{F}{W}\right)$$

F = Frictional resistance
W = Weight

IX. Drainage

A. Hydrology (e.g., Rational method, hydrographs, SCS/NRCS method), including runoff detention/retention/water quality mitigation measures

The rational method can be used to determine the flow rate from runoff of a drainage area. The equation is:

Q = ACi

Q = Flow Rate (cfs)
A = Drainage Area (Acres)
C = Runoff Coefficient
i = Rainfall Intensity (in/hr)

NRCS/SCS Runoff Method

This is an alternative method for determining runoff:

$$S = \frac{1000}{CN} - 10$$

S = Storage Capacity of Soil (in.)
CN = NRCS Curve Number

$$Q = \frac{\left(P_g - 0.2S\right)^2}{P_g + 0.8S}$$

Q = Runoff (in.)
P$_g$ = Gross Rain Fall (in.)

Hydrograph development and applications, including synthetic hydrographs

Hyetographs – Graphical representation of rainfall distribution over time

Hydrograph – Graphical representation of rate of flow vs time past a given point often in a river, channel, or conduit. The area under the hydrograph curve is the volume for a given time period

Parts of a Hydrograph are shown graphically:

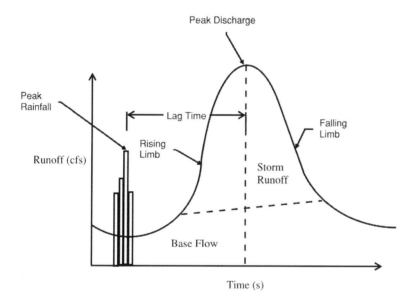

Unit Hydrographs can be determined by dividing the points on the typical hydrograph by the average excess precipitation.

Synthetic Hydrographs are created if there is insufficient data for a watershed. This method uses the NRCS curve number and is a function of the storage capacity.

$$S = \frac{1000}{CN} - 10$$

To develop the synthetic hydrograph, you must calculate the time to peak flow:

$$t_p = 0.5t_R + t_1 \ where \ t_1 = \frac{L_o^{0.8}(S+1)^{0.7}}{1900\sqrt{S_{percentage}}}$$

t_R = Storm duration (time)
L_o = Length overland (ft)
$S_{Percentage}$ = Slope of land

The equation for peak discharge from a synthetic hydrograph then is:

$$Q_p = \frac{0.756 A_{d,Acres}}{t_p}$$

B. Hydraulics, including culvert and stormwater collection system design (e.g., inlet capacities, pipe flow, hydraulic energy dissipation), and open-channel flow

Pressure conduits refer to closed cross sections that are not open to the atmosphere such as pipes:

The Darcy Equation is used for fully turbulent flow to find the head loss due to friction. The equation is:

$h_f = (fLv^2)/(2Dg)$

h_f = Head Loss due to friction (ft)
f = Darcy friction factor
L = Length of pipe (ft)
v = Velocity of flow (ft/sec)
D = Diameter of pipe (ft)
g = Acceleration due to gravity, (Use 32.2 ft/sec^2)

The Hazen-Williams equation is also used to determine head loss due to friction. Be aware of units as this equation may be presented in different forms. The most common is the following:

$h_f = 10.44 L V^{1.85}/C^{1.85} d^{4.87}$

h_f = Head Loss due to F\friction (ft)
L = Length (ft)
V = Velocity (gallons per minute)
C = Roughness coefficient
d = Diameter (in)

Open-channel flow

For open channel flow use the Chezy-Manning equation:

$Q = (1.49/n)AR^{2/3}S^{1/2}$

Q = Flow Rate (cfs)
n = Roughness Coefficient
A = Area of Water (ft^2)
R = Hydraulic Radius (ft)
S = Slope (decimal form)

The hydraulic radius is the area of water divided by the wetted perimeter which is the perimeter of the sides of the channel which are in contact with water.

Hydraulic energy dissipation

A weir is a low dam used to control the flow of water. Weirs have shaped outlets notched into the top of the dam to allow water to flow out. The most common shapes are triangular and trapezoidal:

Triangular Weir

$$Q = C_2 \left(\frac{8}{15} tan \left(\frac{\theta}{2} \right) \right) \sqrt{2g} H^{5/2}$$

$$Q = 2.5H^{5/2} \ (For \ 90^o \ weir)$$

H = Height of water (ft)
θ = Weir angle

Trapezoidal Weir

$$Q = \frac{2}{3} C_d b \sqrt{2g} H^{3/2}$$

Often, the weir can be approximated by taking C_d, the discharge coefficient = 0.63 and the equation is simplified as:

$$Q = 3.367bH^{3/2}$$

63

b = Width of base (ft)

Broad Crested Weirs (Spillways)

Spillways are used to control the flow of excess water from a dam structure. Essentially they are large weirs and therefore can be called broad crested weirs. The calculation of discharge for spillways is taken as:

$$Q = C_s b \left(H + \frac{v^2}{2g} \right)^{3/2}$$

However, often in a dam situation the approach velocity can be taken as zero since it is so small and the equation becomes:

$$Q = C_s b (H)^{3/2}$$

C_s = Spillway coefficient

There are many components used in the collection of stormwater. Some examples include:

Culverts: A pipe carrying water under or through a feature. Culverts often carry brooks or creeks under roadways. Culverts must be designed for large intensity storm events.

Stormwater Inlets: Roadside storm drains which collect water from gutter flow or roadside swales.

Gutter/Street flow: Flow which travels along the length of the street. Gutter flow can be approximated often by an adaptation of the Manning Equation:

$$Q = K(^z/_n)s^{1/2}y^{8/3}$$

K = Gutter flow constant = 0.56 ft³/(s*ft)
z = Inverse of the cross slope of the gutter (Decimal)
n = Roughness coefficient
s = Slope of the gutter (Decimal)
y = water depth at the curb (ft)

Storm Sewer Pipes: Pipes installed under the road which carry the water from inlets to a suitable outlet.

X. Alternatives Analysis

A. Economic analysis (e.g., present worth, lifecycle costs)

The PE exam will potentially provide examples of engineering economics which are geared towards assets in Transportation. For these questions refer to the morning session equations for engineering economics.

MORNING BREADTH PRACTICE EXAMS

Question M1

The chart below gives an estimation of the area of sub-grade to be cut at stations along the baseline of a roadway. Determine most nearly the total volume of excavation in cubic yards from the data.

STATION	AREA OF CROSS SECTION (FT^2)
1+00	0
1+50	155
2+00	170
2+50	65
3+00	0

(A) 722
(B) 19500
(C) 800
(D) 19222

67

Question M2

Given the dimensions of the 8' tall proposed concrete retaining wall shown below determine the area of formwork in square feet required to complete construction of the wall portion only after the footing has been poured.

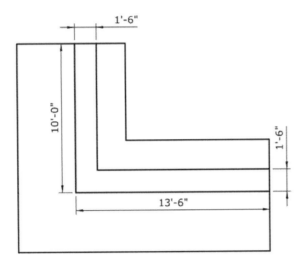

(A) 950
(B) 400
(C) 376
(D) 450

Question M3

Given the scheduling data shown below, determine the length of the critical path in months.

Activity	Predecessor	Duration
A		5 (Months)
B	A	3
C	A	2
D	C	3
E	D	4
F	E	1
G	D	2
H	G	4

(A) 8
(B) 15
(C) 16
(D) 24

Question M4

Design alternatives are being proposed for a bridge replacement with a life span of 30 years. Alternative A has an initial cost of $100,000. It will also require anticipated maintenance costs of 10,000 and 15,000 at years 10 and 20 respectively. Use a rate of 3% and determine most nearly the present worth of the design alternative.

(A) $99050
(B) $110444
(C) $113250
(D) $115746

Question M5

A soil sample has 20% fines and over 50% finer than the no. 4 sieve. It also has a Liquid Limit of 55 and a Plastic Limit of 23. Determine the USCS classification group symbol of this soil.

(A) SC
(B) GC
(C) CL
(D) SM

Question M6

Given the beam and loading conditions shown below, determine most nearly the maximum service moment (kip-ft).

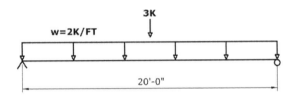

(A) 100
(B) 115
(C) 9.58
(D) 57.5

Question M7

Determine the appropriate effective length factor for the design of a column with end conditions of pinned and pinned.

(A) 1.0
(B) 0.9
(C) 1.2
(D) 0.8

Question M8

Which of the following accurately represents the shear diagram for the beam and loading conditions shown?

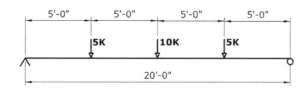

(A)

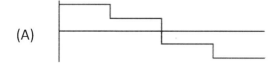

(B)

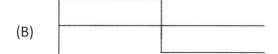

(C)

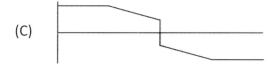

(D)

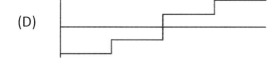

Question M9

Given the beam and loading condition shown below, determine most nearly the moment at point C in Kip-ft.

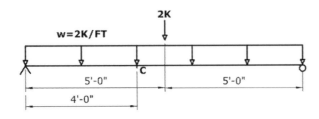

(A) -10
(B) 15
(C) 28
(D) 40

Question M10

Three 6-inch diameter x 12-inch-long concrete cylinders were taken from a concrete batch used to form a retaining wall. The cylinders broke at compressive loads of 102000, 111000, and 100500 pounds. Determine most nearly the average compressive strength of the concrete (psi).

(A) 3200
(B) 4000
(C) 3600
(D) 3700

Question M11

A horizontal curve has a point of curvature at station 1+50 and a point of tangency at station 2+25. If the interior angle of the curve is 10°15', determine most nearly the radius of the curve.

(A) 330'
(B) 420'
(C) 500'
(D) 600'

Question M12

Four open channel concrete cross sections with dimensions and water elevations are shown below. Determine which of the channels has the most efficient section.

(A)

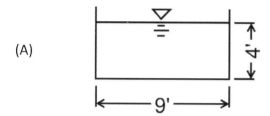

(B)

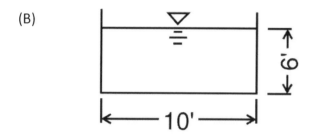

(C)

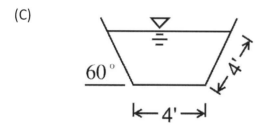

(D)

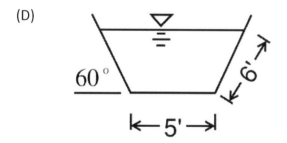

Question M13

Determine the elevation of the roadway at the midpoint of a Vertical Curve with G_1=-2.5% and G_2=2%. The PVC elevation is 260.0' at station 10+00. The Length of Curve is 800'.

(A) 245.0'
(B) 248.6'
(C) 250.0'
(D) 255.0'

Question M14

A horizontal curve is proposed to be constructed around a building. A car traveling along the curve has a line of sight past the building to an object in the road from the beginning of curve to the end of curve. Determine most nearly the minimum distance from the center of road to the building for a required stopping sight distance of 220' to the object if the radius of the curve is 700'.

(A) 8.63'
(B) 15.55'
(C) 20.20'
(D) 35.0'

Question M15

Backfill material is to be transported to a construction site to achieve a proposed grade behind a retaining wall. A total of 100 cubic yards is necessary In the final condition. Determine most nearly in cubic yards how much fill is needed to be taken from the offsite location if there is a swell factor of 1.07, an assumed loss during transport of 5%, and the backfill will be compacted to 90%?

(A) 100
(B) 102
(C) 106
(D) 109

Question M16

A mix design for a parking garage will use 500 lbs of cement per cubic yard. What volume of water in cubic feet per cubic yard should be used to achieve a w/c ratio of 0.65?

(A) 5.21
(B) 6.00
(C) 4.31
(D) 7.50

Question M17

A retaining wall without a batter is 15.5' tall from the bottom of footing and is used to hold granular soils with an angle of internal friction of 25. Determine most nearly the resultant service moment at the base of the footing using Rankine active earth pressure from the soil only. The angle of fill is horizontal. Neglect friction between the wall and the soil and use a soil density of 115 pcf.

(A) 24.7 k-ft
(B) 26.5 k-ft
(C) 28.9 k-ft
(D) 32.1 k-ft

Question M18

A saturated soil sample has a weight of 50 lbs. The sample is placed in an oven and then weighed to measure 42 lbs. The specific gravity of the soil was determined to be 2.4. Determine most nearly the void ratio of the sample.

(A) 0.28
(B) 0.40
(C) 0.42
(D) 0.46

Question M19

A vehicle is traveling at a velocity of 100 ft/s on a 2% incline. If the driver has a 2 second breaking perception reaction time, determine the total distance in feet it takes to stop the vehicle. Assume a coefficient of friction of 0.3.

(A) 200.0
(B) 684.5
(C) 480.9
(D) 700.0

Question M20

Which of the following is added to the tangent runout to calculate the superelevation transition distance?

(A) Superelevation runoff
(B) Cross slope
(C) Approach Grade
(D) Degree of Curvature

Question M21

Assuming fully turbulent flow, determine most nearly the velocity in ft/sec for a pipe with a diameter of 2 ft, a length of 500 ft and a head loss due to friction of 3 ft. Use a Darcy friction factor 0.02.

(A) 1.0
(B) 5.0
(C) 6.2
(D) 11.4

Question M22

A rectangular open channel is 6 ft wide and has a water depth of 3 ft. Determine most nearly the flow rate (cfs). The channel has a roughness coefficient of 0.015 and a slope of 0.003.

(A) 130
(B) 138
(C) 145
(D) 200

Question M23

For a head loss due to friction of 40 ft, determine most nearly the volumetric flow rate in gallons per min for a pipe with length 100' and diameter of 0.25 ft. Use a Hazen-Williams coefficient of 140.

(A) 200
(B) 265
(C) 410
(D) 432

Question M24

Which of the following defines the time base for a stream hydrograph?

(A) Time from the base flow until the peak flow
(B) Time from the peak flow until flow drops below the base flow
(C) Time that the flow exceeds the base flow
(D) Time that the flow drops below the base flow

Question M25

A circular culvert is to be designed to handle the runoff from two areas of drainage. The first has an area of 15 Acres, a runoff coefficient of 0.18, and a rainfall intensity of 1.5 in/hr. The second has an area of 10 Acres, a runoff coefficient of 0.22, and a rainfall intensity of 1.5 in/hr as well. Determine most nearly the minimum culvert area in square feet to limit the flow velocity to 0.26 ft/s.

(A) 28.28
(B) 29.50
(C) 31.0
(D) 62.0

Question M26

Given the truss configuration shown below, how many 0-Force members are there? The truss is simply supported and points A and H.

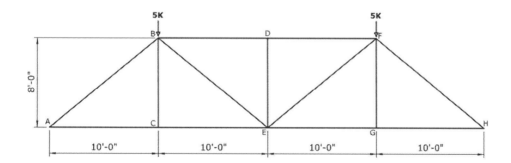

(A) 1
(B) 2
(C) 3
(D) 5

Question M27

Given the truss configuration shown below, what is the support reaction at node A? Neglect the self-weight of the truss. The truss is simply supported and points A and H.

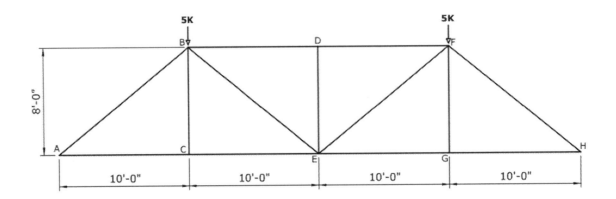

(A) 2 Kips
(B) 5 Kips
(C) 6.67 Kips
(D) 10 Kips

Question M28

Given the truss and loading condition in question M27, what is the axial force in member C-E? Neglect the self-weight of the truss.

(A) 0 Kips
(B) 5 Kips
(C) 6.24 Kips
(D) 8.71 Kips

Question M29

Which of the following is most often the cause of rebar corrosion?

(A) Chloride intrusion of concrete
(B) Freeze-thaw action
(C) Delayed Ettringite Formation
(D) High water to cement ratio

Question M30

Which of the following mix design properties has the greatest impact on the strength of the concrete?

(A) w/c ratio
(B) Percent Coarse Aggregate
(C) Void Ratio
(D) Percent Fine Aggregate

Question M31

Water exits a reservoir through a pipe 50' below the water surface. Determine the velocity of flow at the exit of the pipe. Assume frictionless flow and the discharge is at atmospheric pressure.

(A) 45.66 ft/s
(B) 56.75 ft/s
(C) 60.10 ft/s
(D) 75.55 ft/s

Question M32

Which of the following is not an assumption of the Bernoulli energy conservation equation?

(A) The fluid is incompressible
(B) There is no fluid friction
(C) Changes in thermal energy are negligible
(D) The potential energy is zero

Question M33

Flow from pipes A and B connect to exit out a single pipe C and have velocities of 1.2 and 0.8 ft/s respectively. If pipe A has an area of 3.0 ft² and pipe B an area of 4.0 ft², what is the flow rate for pipe C in cubic feet per second?

(A) 3.6
(B) 4.0
(C) 6.0
(D) 6.8

Question M34

After an onsite investigation, the soil profile below has been developed from borings. Determine most nearly the effective stress at a depth of 40 ft for the saturated soil densities indicated.

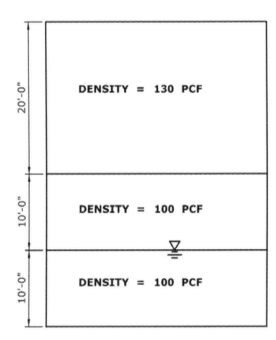

(A) 2600 psf
(B) 3550 psf
(C) 3975 psf
(D) 4600 psf

Question M35

A simply supported 4" wide X 6" deep beam is 15' long and has two point loads of 10 kips applied at 5' and 10' from the left end, what is the maximum bending stress of the cross section?

(A) 25 ksi
(B) 35 ksi
(C) 50 ksi
(D) 55.5 ksi

Question M36

A retaining wall is to be designed by limiting the shear stress in the stem to 0.1 ksi/ft. As shown below the horizontal force due to earth pressure is calculated as 9 Kips. The retaining wall has a batter of 8:1. Determine most nearly the minimum width of the base of the stem, b, if the depth to the flexural reinforcement is 6".

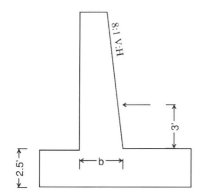

(A) 6.0"
(B) 7.5"
(C) 8.25"
(D) 12"

Question M37

Which of the following roadside safety barriers is most appropriate for an object which is 2'-6" perpendicular from the edge of roadway?

(A) Concrete Barrier
(B) 3-Cable Guiderail System
(C) Metal Beam Rail
(D) None

Question M38

For the traffic counts shown below, determine the peak hourly traffic volume.

Time	Volume
8:00-8:15	500
8:15-8:30	560
8:30-8:45	650
8:45-9:00	625
9:00-9:15	630
9:15-9:30	600
9:30-9:45	540
9:45-10:00	460

(A) 2260
(B) 2455
(C) 2505
(D) 2600

Question M39

Given a 6" wide X 12" deep cantilever beam with a point load at the free end of 1 K, determine the maximum deflection on the beam for a length of 10'. Use a modulus of elasticity of 3605 ksi.

(A) 0.10"
(B) 0.18"
(C) 0.28"
(D) 0.50"

Question M40

Determine the specific gravity of a soil sample. The total volume is 2 cubic ft and the volume of the soil is 1.5 cubic ft. The degree a saturation is 75% and the moisture content is 0.1.

(A) 2.2
(B) 2.5
(C) 2.8
(D) 2.95

Question M41

A soil sample is analyzed in the lab and it is determined to have 32% fines. The Liquid Limit is 36 and the Plastic Limit is 24. Determine the AASHTO Soil Classification of the sample.

(A) A-1-a
(B) A-2-4
(C) A-2-6
(D) A-4

Question M42

Charpy V-Notch is a test for which property of steel?

(A) Durability
(B) Strength
(C) Ductility
(D) Toughness

Question M43

Loads are determined for the design of a third story building. The types and magnitude of the loads are as follows:

Floor Slab Self-Weight = 100 psf
Flooring = 10 psf
Pipes and Utilities = 2 psf
Pedestrian Loading = 80 psf
Furniture = 40 psf

Using the Load combination of 1.2DL + 1.6LL, determine the design distributed load.

(A) 232.0 psf
(B) 280.6 psf
(C) 326.4 psf
(D) 380.0 psf

Question M44

A concrete beam is 6" wide x 12" deep and has a uniform distributed load of 2 k/in as shown below. Determine most nearly the maximum deflection of the beam using a modulus of elasticity of 3605 ksi.

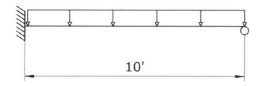

10'

(A) 0.25"
(B) 0.55"
(C) 0.68"
(D) 0.72"

Question M45

A state department of transportation initiates a project to construct a maintenance facility for the housing of snow plows. The footprint of the building falls within state property however there is a need to store materials in an area past the property line of an adjacent privately-owned facility. Which of the following is the most appropriate Rights of Way action to complete construction?

(A) Temporary Easement
(B) Permanent Easement
(C) Property Acquisition
(D) No action required

Question M46

The cross section for a 20' long retaining wall is shown below. The dimensions of the cross section are consistent throughout the length. A chart is developed for the items identified for the completion of the work and unit prices have been developed using previous projects. Determine the cost of the work for the identified items.

Item	Unit	Unit Price
Concrete Wall	Cubic ft.	30
Backfill	Cubic Yd.	100
6" Dia. Drain	Linear ft.	20

(A) 15007
(B) 22104
(C) 33550
(D) 55250

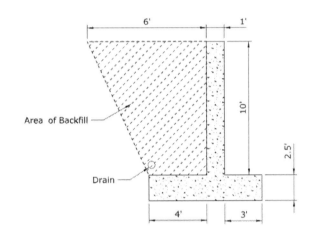

Question M47

A multi-use trail is to be constructed by crossing under an existing roadway using precast box culvert sections and detouring traffic. Which of the following is an appropriate sequence for the construction activities?

(A) Close roadway to traffic, place box culvert, backfill, pave road and open to traffic
(B) Set up detour signalization, close roadway to traffic, excavate, place box culvert, backfill, pave road and open to traffic
(C) Set up detour signalization, close roadway to traffic, excavate, place box culvert, allow for curing time, backfill, pave road and open to traffic
(D) Set up detour signalization, excavate, close roadway to traffic, place box culvert, backfill, pave road and open to traffic

Question M48

The chart below shows lift capacities (in pounds x 1000) of a crawler crane with a 142,000 lb counter weight. Using a boom length of 120' determine most nearly the maximum swing radius when lifting a structural member weighing 75 tons.

Boom(ft)	100	120	140
Radius (ft)			
24	191.7	175.9	157.9
30	160.2	152.2	140.6
34	136.0	134.7	127.5
40	107.8	107.9	107.7

(A) 24'
(B) 30'
(C) 34'
(D) 40'

Question M49

A bridge with approximately an 80' span is to be replaced in an urban area with high average daily traffic numbers. It has been determined that it is acceptable to detour the road for short period of time to allow for the replacement of the bridge. Which of the following is the most appropriate construction method?

(A) Cast in place reinforced concrete
(B) Precast concrete
(C) Prestressed-precast concrete
(D) Post-tensioned concrete

Question M50

A temporary shoring tower is being used to support a 30' long steel girder beam as shown below. Determine most nearly the minimum distance L which will limit the unfactored design reaction in the tower to a maximum of 5 kips. The steel beam has a self-weight of 288 lb/ft.

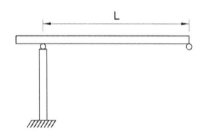

(A) 18'
(B) 22'
(C) 26'
(D) 29'

Question M51

Which of the following excavation scenarios would require the use of a temporary earth retaining systems?

(A) 8' deep cut into sandy soil
(B) 4' deep cut into clay
(C) 10' deep cut in stable rock
(D) 6' deep cut into stable rock

88

Question M52

An open cut excavation takes place to facilitate the placement of concrete pipes. The excavation is sloped to prevent failure at a rate of 1.5:1.0 V:H. A boring log indicates the soil is saturated clay with a density of 125 pcf and a cohesion of 300 psf. What is the maximum length of flat workspace for a factor of safety of 1.5 if the available right of way is 100' long? Use a slope stability number of 6.2.

(A) 20.0'
(B) 61.0'
(C) 73.6'
(D) 90.5'

Question M53

A pipe as shown below handles water with a flow rate of 2.0 cubic feet per second. Determine the minor head loss between point A and B. The K factor for a 90° bend was determined as 0.9.

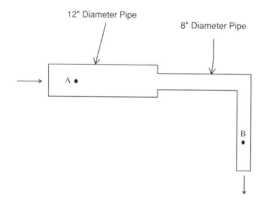

(A) 0.12'
(B) 0.46'
(C) 1.20'
(D) 2.29'

Question M54

A 25' long 3' x 3' concrete column is constructed to support a concentrated load of 50 kips. The column is fixed at the base and free to rotate but not translate at the top. Determine the Euler critical stress if the Modulus of Elasticity is 3605 ksi.

(A) 18.2 ksi
(B) 45.0 ksi
(C) 58.8 ksi
(D) 66.8 ksi

Question M55

Which of the following USCS Group Symbol soil types would be expected to undergo the most primary consolidation?

(A) GW
(B) GP
(C) SP
(D) CH

Question M56

Which of the following appropriately defines a design 100-year storm?

(A) A storm that occurs once per year over a span of 100 years
(B) A storm event with the 100th greatest intensity in a given year
(C) A storm with an intensity which will only occur once in 100 years
(D) A storm with an intensity of 100

Question M57

A detention pond is to be designed to limit the flow of runoff at the base of a slope. Water exits the pond by a 2' diameter pipe and it is desired to limit the flow velocity to 0.25 feet per second. Determine most nearly the minimum size of the pond in cubic yards to prevent overtopping of the pond for a 6-hr storm event with an average runoff flow rate into the pond of 0.85 cubic feet per second.

(A) 25
(B) 52
(C) 75
(D) 150

Question M58

A soil sample is taken using a split spoon method with a 140 lb hammer. The chart below shows the number of blows for each 6" increment. Determine most nearly the average N value of the soil sample for the depth shown.

Depth (in)	Blows per 6" Increment
6	5
12	10
18	16
24	18
30	20
36	28
42	32

(A) 22
(B) 28
(C) 35
(D) 41

Question M59

A permeability test is performed on a cylindrical soil sample with an approximate diameter of 6" and a length of 12". It was measured that 1 cubic in of water took 10 seconds to enter and exit the sample. The difference in pressure head was measured as 3". Determine most nearly the hydraulic conductivity in in/s of the sample.

(A) 0.005
(B) 0.014
(C) 0.110
(D) 1.140

Question M60

During the construction of a bridge replacement project bearing pads are surveyed to verify they are set at the appropriate elevation. The first measurement is recorded as 158.5682'. Which of the following is the appropriate recorded measurement with the appropriate level of accuracy?

(A) 158.568
(B) 158.57
(C) 158.6
(D) 159.0

Question M61

Which of the following methods is most often used to permanently control sediment from runoff of excavated areas?

(A) Silt Fence
(B) Hay Bales
(C) Seeding and Turf Establishment
(D) Erosion Control Fabric

Question M62

A 14" x 14" concrete column fixed at the base is subjected to an eccentric load of 5 kips as shown below. Determine most nearly the maximum compressive stress in the column at the base. Ignore self-weight of the column.

(A) 0.091 ksi
(B) 0.140 ksi
(C) 0.255 ksi
(D) 0.820 ksi

Question M63

A project is determined to need 14' x 10' box culverts of varying lengths in 3 different locations. Which of the following is the most appropriate unit of measure for the estimation of box culvert item?

(A) Linear ft.
(B) Square ft.
(C) Each
(D) Lump Sum

Question M64

A 10' X 10' spread footing is shown below, determine ultimate moment at the critical section for the design of flexural reinforcement. The factored point load includes the self-weight of the column and footing and is 50 K.

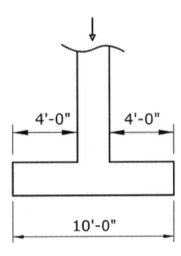

(A) 10 K - ft
(B) 18 K - ft
(C) 25 K - ft
(D) 40 K - ft

Question M65

Determine most nearly the time of concentration for sheet flow of a watershed which has a distance to the outlet of 250'. The watershed is short prairie grass which has a Manning's Roughness Coefficient of 0.15 and has a slope of 0.05 ft per ft. The 2 yr, 24 hr rainfall is given as 2.2 inches.

(A) 0.284 hrs.
(B) 0.313 hrs.
(C) 0.385 hrs.
(D) 0.486 hrs.

Question M66

Traffic data from an observation of highway is shown below. Determine most nearly the peak hour factor.

Time Interval	Volume(# of vehicles)
7:00-7:15	1900
7:15-7:30	2200
7:30-7:45	2350
7:45-8:00	2000
8:00-8:15	1700
8:15-8:30	1200

(A) 0.85
(B) 0.87
(C) 0.90
(D) 0.95

Question M67

A storm event produces an average depth of 3.5 inches. Using an NRCS curve number of 79, determine the direct runoff.

(A) 1.56"
(B) 1.78"
(C) 2.5"
(D) 4.4"

Question M68

A reinforced concrete wall is supported on a strip footing 8' wide and 2.5' deep from the soil surface to the bottom of the footing. The soil below is cohesion less, has a unit weight of 125 pcf, a density bearing capacity factor of 2.5 and a depth bearing capacity factor of 4.4. Determine the allowable bearing capacity of the soil for a safety factor of 2.5.

(A) 0.925 ksf
(B) 1.650 ksf
(C) 2.255 ksf
(D) 2.58 ksf

Question M69

Which of the following cement types is most appropriate for the construction of a mat foundation where sulfate attack is not a concern?

(A) Type I
(B) Type II
(C) Type III
(D) Type IV

Question M70

During a 3 hr storm event, a hydrograph is developed as shown below by taking measurements at 1-hour intervals of the flow rate. The storm intensities in half hour increments were also recorded and provided in the table below. Determine most nearly using a histographic approximation of the hydrograph the total volume of water in cubic feet during the lag time of the storm event.

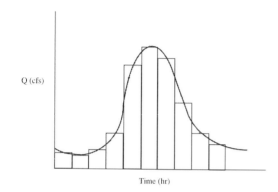

	A	B	C	D	E	F	G	H	I	J
Flow Rate (cfs)	1.1	0.5	1.4	2.5	6.6	8.6	6.9	4.5	2.6	1.6
Time (hrs)	1	2	3	4	5	6	7	8	9	10

Intensity (in/hr)	0.2	0.5	1.5	1.8	0.4	0.1
Time (hrs)	0.5	1	1.5	2.0	2.5	3.0

(A) 1857
(B) 54180
(C) 146160
(D) 211680

Question M71

An 8' wide x 8" deep precast wall panel is to be erected by being lifted from a flat position as shown below. The self-weight is 0.8 K/ft. Determine the minimum strength of concrete, f'$_c$, to prevent tensile cracking during lifting. Assume the wall panel can be analyzed as a simply supported beam between the pick points. Do not incorporate any safety factors or load factors.

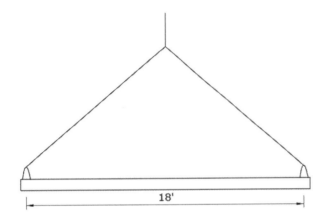

(A) 2250 psi
(B) 2480 psi
(C) 2570 psi
(D) 3200 psi

Question M72

A department of transportation is trying to determine its fiscal budget for the next five years. Three projects are being evaluated specifically which have a proposed cost of 100000, 175000, and 225000 and are scheduled to be paid for 3 years, 4 years, and 5 years from today respectively. Using an inflation of 3%, determine the total cost to the department.

(A) 475000
(B) 500000
(C) 525090
(D) 567073

Question M73

Which of the following is the least appropriate shape for the design of flexure about the X-axis?

(A)

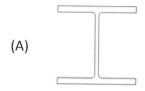

(B)

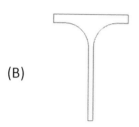

(C)

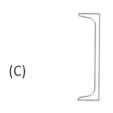

(D)

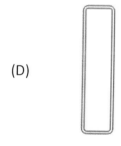

Question M74

Which of the following is not a method for the protection of steel against corrosion?

(A) Weathering Steel
(B) Galvanized Steel
(C) High Strength Steel
(D) Painting Systems

Question M75

A bridge is to be replaced which carries a local road over a highway. A boring log reveals the soil types for the construction of the north east wingwall of the bridge to be sandy soils to a depth of 10 ft. and then bedrock 10 ft. and below. The wingwalls are anticipated to be about 10' high. Which of the following is the most appropriate foundation type for the wingwall?

(A) Spread Footing
(B) Micro-pile
(C) Pile
(D) Drilled Shaft

Question M76

A project schedule is shown below with tasks designated by letters and durations above each task in days to complete for a single crew. This project includes two separate crews of workers and the crews can work either separately or simultaneously on any task at any time. The cost per hour for workers is $50/worker and crews are made up of 4 workers each. Work days are 8 hrs. long. Determine the minimum cost of labor to complete the project.

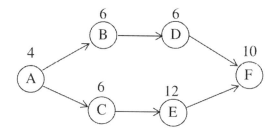

(A) 5250
(B) 22200
(C) 35200
(D) 100500

Question M77

A project schedule is determined and an activity diagram is developed as shown below. Predecessors are indicated with arrows and the associated original durations are indicated in days by the numbers. However, due to competing priorities, the durations of tasks C, D, and F need to increase by 1, 4, and 3 days respectively. Determine the new duration in days for the critical path of the revised schedule.

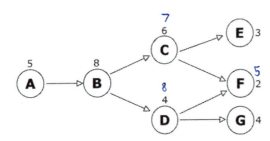

(A) 22
(B) 23
(C) 25
(D) 26

Question M78

An 8" thick one-way slab spans 20' between wall panels and is 10' wide. The slab is fixed at the connection to the walls. Including self-weight, the slab is to be designed for a distributed load of 0.5 ksf. Determine most nearly the un-factored design moment in the design of the slab.

(A) 9.50 k-ft
(B) 12.55 k-ft
(C) 14.20 k-ft
(D) 16.67 k-ft

Question M79

The roadway cross section below includes a construction stake with the markings shown.
Determine the elevation of the top of curb if the height of the curb from the roadway is 10".

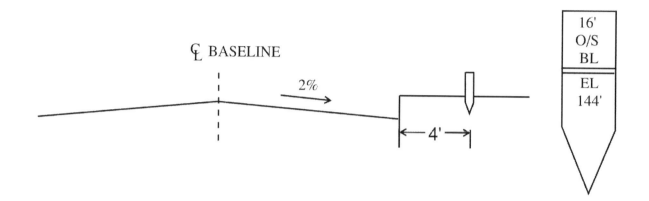

(A) 143.75'
(B) 144.0'
(C) 144.60'
(D) 145.8"

Question M80

A tensile lab test is performed on a cylindrical steel member having a diameter of 0.25". From
the results below determine most nearly the yield strength (ksi) of the steel.

Load (lbs)	437	778	1502	2033	2262	2656	2428	2392	2601	3180	3297
Elongation (in.)	0.019	0.021	0.025	0.027	0.028	0.03	0.031	0.059	0.098	0.23	0.32

(A) 36
(B) 45
(C) 54
(D) 60

TRANSPORTATION DEPTH PRACTICE EXAM

Question T1

A field observation of traffic speeds and volumes on a 3-lane freeway with level terrain is performed over a 0.75 mile stretch. The peak hour volume is determined to be 1400 vph with a peak hour factor of 0.93. The speed observations for 10 vehicles are shown below and are determined to be representative of the average speed. The freeway has 12' lanes and 13' shoulders with 5% truck traffic. Determine the level of service of the freeway.

Vehicle	Speed (sec)
1	34
2	43
3	44
4	48
5	52
6	31
7	36
8	37
9	45
10	48

(A) A
(B) B
(C) C
(D) E

Question T2

A cross section of a roadway along a horizontal curve with a radius of 1500' is shown below. It is determined that the Average Annual Daily Traffic can be taken as 1250 pcpd and the design speed is 45 mph. Determine the minimum width of cross section (X) necessary to have an acceptable roadway for roadside clearance.

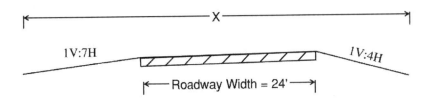

(A) 16.5'
(B) 38.0'
(C) 54.6'
(D) 57.2'

Question T3

A sidewalk in plan view is shown below. Determine the effective walkway width using the maximum of the recommended range values in the Highway Capacity Manual. Use a shy distance from the store front of 1.5'.

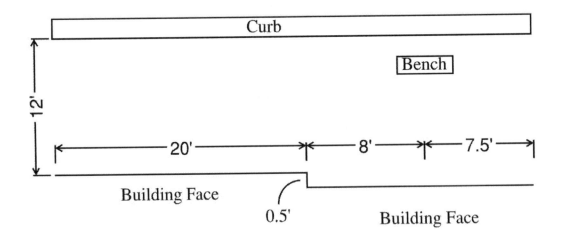

(A) 4.0'
(B) 4.5'
(C) 5.5'
(D) 6.5'

Question T4

Data from a generation study involving trips from residential buildings based on the number of units is approximated by the four data points shown below. A new building with 450 units is proposed to be built in the same area. Determine the anticipated number of trips resulting from the new building.

# of dwelling units	trips
200	520
275	670
325	770
390	900

(A) 850
(B) 920
(C) 990
(D) 1020

Question T5

A through lane for an intersection has a base flow rate of 1900 pcphpl and an existing calculated saturation flow rate of 2500. A proposed project for the intersection will increase the lane width from 9.5′ to 12′. Determine the saturation flow rate (pcphpl) after the intersection changes are complete.

(A) 2443
(B) 2605
(C) 2765
(D) 2900

Question T6

A 9-mile long stretch of a roadway has a AADT of 1250 and includes three on ramps as shown below. There is a proposal for modifications to the on ramps with current and predicted crash frequencies per year as shown in the diagram. The State DOT provides the following predictive model for the Safety Performance Function crash frequency per year:

$$N_{SPF} = Le^{-1.2}(AADT)^{0.5}$$

Determine most nearly the predicted crash frequency of the stretch of roadway per year using a calibration factor of 1.1.

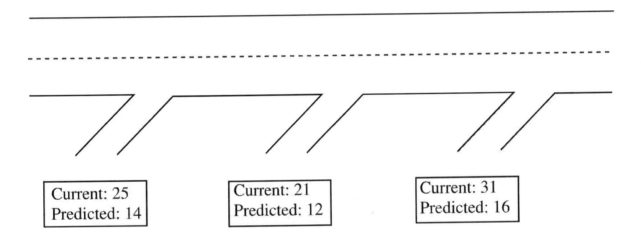

Current: 25
Predicted: 14

Current: 21
Predicted: 12

Current: 31
Predicted: 16

(A) 9
(B) 18
(C) 24
(D) 35

Question T7

A crest vertical curve ends at station 25+00. The middle ordinate of the curve needs to be held to 3 ft. Determine the station at the beginning of the curve to maintain the stopping sight distance of 155'.

(A) 22+18
(B) 23+02
(C) 23+37
(D) 24+22

Question T8

A vehicle traveling at 60 mph needs to maneuver away from an accident that just occurred. The sum of the pre-maneuver and the maneuver times is 12.4 seconds. Determine the type of avoidance maneuver as per the *Policy on the Geometric Design of Highways and Streets*.

(A) A
(B) C
(C) D
(D) E

Question T9

A 3-Lane freeway with 12' lanes will have a PVC at station 45+00. For a superelevation rate of 7%, determine the station at which the cross section is fully superelevated if the design speed is 60 mph.

(A) 45+75
(B) 46+25
(C) 49+75
(D) 50+82

Question T10

A crest vertical curve has G_1 = 2%, G_2 = -4% and a PVI elevation of 155.0' at station 10+00. A bridge has a low point elevation of 170' to the bottom of the girder at station 8+00. Determine most nearly the vertical clearance between the road and low point of the bridge if the length of curve is 800'.

(A) 17.0'
(B) 20.5'
(C) 26.0'
(D) 29.5'

Question T11

A 10 mile stretch of a divided 4-lane highway is to be rehabilitated. The Highway has 12' lanes and the lateral clearances are 6' on the right and 4' on the left. There are currently 45 access points along the stretch which can be classified as an interruption to flow. The rehab project will eliminate 5 of the access points. Determine the design free flow speed for the rehabilitation project if the base flow can be taken as 65 mph.

(A) 45.5 mph
(B) 56.0 mph
(C) 63.6 mph
(D) 72.2 mph

Question T12

A 5-mile stretch of roadway is evaluated over a three-year period for crashes. The accidents recorded are 28, 31, and 45 for each year. The average daily traffic can be taken as 1200 vehicles per day for the east direction and 1600 vehicles per day in the west direction. Determine most nearly the accident rate per 100 million vehicle miles.

(A) 56
(B) 220
(C) 532
(D) 678

Question T13

Determine the station of the low point of a Vertical Curve with G_1=-3% and G_2=2%. The PVI elevation is 150.0' at station 10+00. The Length of Curve is 600'.

(A) 9+00
(B) 9+50
(C) 10+00
(D) 10+60

Question T14

Repairs to a concrete curb on the side of a 2-Lane urban roadway will require a temporary work zone. The proposed traffic control will be a merging pattern with a width of 10' needed. If the design speed for the roadway is 25 mph, which of the following is not a correct requirement for the work zone?

(A) The maximum spacing for channelization devices is 25'
(B) The minimum taper length is about 104'
(C) The minimum distance between advance warning signs is 100'
(D) The advance warning area will require 4 signals

Question T15

Traffic for one direction is analyzed for a 3-Lane highway with 12' lanes and sufficient lateral clearance. The volume is recorded as 2050 pcph. The peak volume in 15 min measured intervals was recorded as 620 passenger cars. This highway does not allow the use of trucks or recreational vehicles. Determine the Level of service for the highway if the measured free flow speed is 60 mph.

(A) A
(B) B
(C) C
(D) D

Question T16

A freeway has two lanes in each direction and enters a horizontal curve with a radius along the centerline of the first interior lane of 4500' and an angle of 12°. The lanes are 12' each and the median is 6' wide. The curve wraps around a building which is offset 30' from the center of the roadway at the midpoint of the curve. Determine the number of lanes which have an obstructed line of sight from the building.

(A) 1
(B) 2
(C) 3
(D) 4

Question T17

A new highway sign will be constructed at station 21+40 which will fall in the clear zone of the roadway. The highway has an average daily traffic volume of 7240 pcpd and a design speed of 60 mph. A metal beam rail will be installed to protect the new highway sign. The distance to the barrier will be the shy distance plus 2 additional feet. If the distance from the edge of roadway to the back edge of sign is 16 feet, determine the station of the length of need for a non-flared barrier.

(A) 18+90
(B) 20+46
(C) 20+90
(D) 21+10

Question T18

A horizontal curve is to be designed for a roadway with a 5% rate of superelevation and a radius of 750 ft. Determine the maximum design speed for the curve to achieve a skid resistance number of 35.

(A) 48 mph
(B) 55 mph
(C) 67 mph
(D) 75 mph

Question T19

A four-way intersection is 2-way stop controlled. The major roadway has two lanes and the minor has one lane. Which of the following statements does not satisfy the requirements of a traffic warrant?

(A) The peak hour ratio of major street vehicles (total of both approaches) to pedestrians is 900:500
(B) The 8-hour major (total of both approaches) to minor (higher approach only) average volume is 600:200
(C) The 4-hour major (total of both approaches) to minor (higher approach only) average volume is 700:150
(D) The peak hour volume of the major street (total of both approaches) to minor (higher approach only) is 900:400

Question T20

Which of the following statements warrant the consideration of the use of a stop sign on the minor street approach of an intersection?

(A) The average daily volume of the minor street is 1254 pcpd
(B) The average daily volume on the major roadway is 6050 pcpd
(C) The crash history indicates two right angle collisions in the past 12 months
(D) The crash history indicates 4 right angle collisions in the past 24 months

Question T21

A 3-lane freeway has a current average daily traffic of 5500. The peak hour factor is given as 0.93 and the freeway sees no trucks or recreational vehicles. Using a design speed of 65 mph, determine the minimum growth rate until the forecasted traffic makes the level of service an F after 15 years.

(A) 2.67%
(B) 3.56%
(C) 4.55%
(D) 5.75%

Question T22

A crest vertical curve has an approach grade of 2% and an exit grade of -3%. The 85th percentile speed for the roadway is 35 mph. Determine the minimum curve length necessary to avoid the need for no passing lane markings

(A) 540'
(B) 578'
(C) 620'
(D) 780'

Question T23

A truck approaches a sag vertical curve with slopes -2% and 2% on a 2-lane roadway with 12' lane widths. The curve crosses an overpass with a cross section as shown below. For a length of curve of 1200' and a design sight distance of 1250', determine the horizontal distance from the face of the abutment to the point which provides proper vertical clearance. Use a height to the driver's eye of 8 ft and a height of the approaching car's headlights of 2 ft.

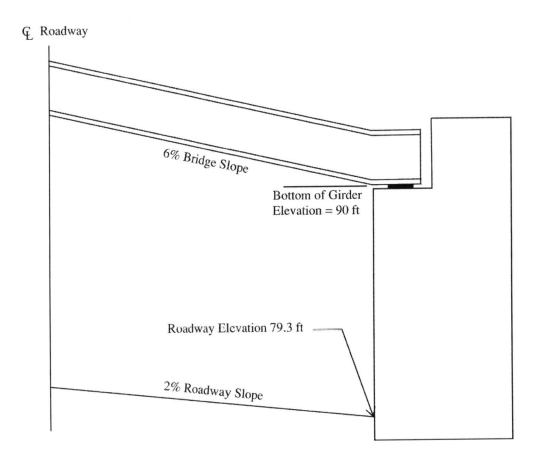

(A) 2'
(B) 8'
(C) 15'
(D) 20'

Question T24

A driver in a 4000 lb vehicle falls asleep at the wheel traveling 45 mph. He veers off into a crash attenuation system of barrels each weighing 2000 lbs. The design requirement calls for the car to come to a stop 15' from impact. Determine the number of barrels needed to stop the vehicle in the allotted distance for a safety factor of 1.2 and an efficiency factor of 0.85.

(A) 4
(B) 6
(C) 10
(D) 13

Question T25

Determine the minimum width of space needed along a horizontal curve to allow an SU-12 design vehicle to complete the minimum turning path without obstruction.

(A) 12.0'
(B) 14.6'
(C) 16.4'
(D) 22.4'

Question T26

A signalized intersection is to be designed for a 2-way, 2-lane roadway with a 3.25' median. A 10' wide cross walk is necessary running North to South on the eastern side of the intersection to handle 15 pedestrians/hour with an average speed of 2 ft/s. The signals have 4 phases totaling 190 seconds. The time in which the signals allow traffic through the east cross walk is 40 seconds. Determine the maximum lane width for the roadway to allow the time to clear to be 25 seconds.

(A) 10'
(B) 11'
(C) 11.5'
(D) 12'

Question T27

A signalized intersection has three phases. The effective green and yellow times for each phase are as shown below. The design of the intersection uses a green reaction time of 1.5 s and a red clearance time of 2 seconds for each phase. Determine the cycle length for the intersection.

Phase	Effective Green (Includes 2 sec extension)	Yellow
1	10	4
2	30	5
3	20	5

(A) 36.5 s
(B) 55.5 s
(C) 60.0 s
(D) 78.5 s

Question T28

Which of the following interchange types is the most appropriate for a freeway intersecting a local road in an urban setting with no influence from adjacent interchanges?

(A) Diamond
(B) Partial Cloverleaf
(C) Full Cloverleaf
(D) Fully Directional

Question T29

Two vehicles approach an intersection as shown below. The distances to the drivers' eyes are dimensioned. The vehicle on the major road is traveling at 55 mph and the minor road is stop controlled. Determine the minimum sight distance at the intersection if the friction factor can be taken as 0.35 and the grade is 2%.

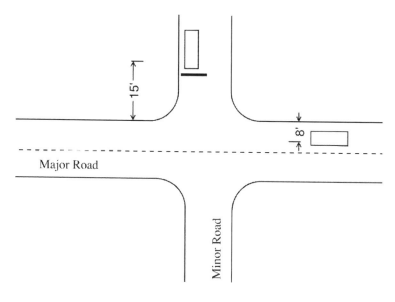

(A) 270.2'
(B) 273.5'
(C) 280.7'
(D) 320.6'

Question T30

A compound horizontal curve consisting of two separate curves is necessary to connect the roadway from station 25+50 to 32+45. The first curve has a radius of 2000' and an angle of 10°. If the second curve has a radius of 2500', determine the required angle of the second curve for the required length.

(A) 6.2°
(B) 7.2°
(C) 7.9°
(D) 11.3°

Question T31

As per the *2010 ADA Standards for Accessible Design,* determine the minimum width needed to provide an acceptable passenger loading zone with a vehicle pull up space and access aisle.

(A) 55"
(B) 60"
(C) 96"
(D) 156"

Question T32

Bituminous pavement is to be designed for two layers over the subgrade: the pavement and the base layer. The pavement has a structural number of 1.8 and the base and pavement combined have a structural number of 4.6. The strength coefficients are 0.42 and 0.11 for the pavement and base. If the base has a drainage coefficient of 0.9, determine the ratio of the required pavement thickness to base thickness.

(A) 0.15
(B) 0.22
(C) 0.34
(D) 0.50

Question T33

Which of the following is the maximum flow rate for a roundabout with a single 12' circulatory lane and four points of entry?

(A) 1800 vph
(B) 2200 vph
(C) 3000 vph
(D) 3600 vph

Question T34

For a 2-mile segment of a central business district urban roadway, the running time is computed as 420 seconds. Determine the transit travel speed if the signal condition is typical with no right turns.

(A) 12.8 mph
(B) 22.5 mph
(C) 34.6 mph
(D) 45.0 mph

Question T35

Drainage for a road with a concrete gutter (Roughness coefficient = 0.012) is to be designed using a slope of 0.3%. The cross slope of the gutter is 2.5% and the height of water is limited to 3". Determine most nearly the design flow rate for the gutter.

(A) 2.53 cfs
(B) 4.56 cfs
(C) 12.22 cfs
(D) 25.34 cfs

Question T36

Which of the following statements are false concerning the design of flexible pavement?

(A) A decrease in the soil resilient modulus will increase the required pavement thickness
(B) CBR values are applicable to soils with a resilient modulus up to 45,000 psi
(C) CBR and R Soil Values are applicable to USCS Soil classes ML, SC, and SP
(D) A CBR value of about 2.9 is equivalent to an R value of about 6

Question T37

Which of the following is the minimum spacing for interchanges on rural highways?

(A) ½ Mile
(B) 1 Mile
(C) 1 ¼ Mile
(D) 2 Miles

Question T38
A roadway is to undergo a profile reconstruction to create a more gradual slope. The profile diagram for the reconstruction is shown below starting at station 14+50. Which of the following stations is nearest to the balancing point for the cut and fill?

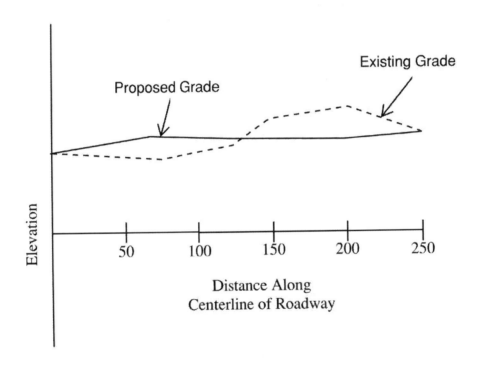

(A) 15+25
(B) 15+80
(C) 16+00
(D) 16+50

Question T39

Three drainage areas have the characteristics as shown below. A retention pond is designed to handle the runoff from these areas. The retention pond has a maximum capacity of 200,000 ft³ and has an average standing volume of 80,000 ft³. Determine the maximum average storm intensity the retention pond can handle without overtopping in 6 hrs. Ignore effects from evaporation and transpiration.

Area 1

5 acres
Coefficient = 0.16

Area 2

4 Acres
0.23

Area 3

6 Acres
0.22

(A) 0.93 in/hr
(B) 1.83 in/hr
(C) 2.01 in/hr
(D) 3.45 in/hr

Question T40

The State Department of Transportation is performing a cost alternative analysis for its highway paving program. A 25-mile stretch of roadway has two proposed alternates for a 30-year lifecycle. The first is milling and paving at a cost of $300,000 every 10 years including the first to be performed at year zero. The second alternative is a full depth reconstruction with an initial cost of $650,000 and a single mill and pave after 20 years of $300,000. Using present worth analysis, determine the cost of the less expensive option using a rate of 3.5%.

(A) $770,400
(B) $800,770
(C) $895,800
(D) $950,000

SOLUTIONS

Solution M1

Use the average end area method

$V = L(A_1+A_2)/2$ where the length between each station is 50'

1+00 to 1+50:

$V = 50(0+155)/2 = 3875$ ft^3

1+50 to 2+00:

$V = 50(155+170)/2 = 8125$ ft^3

2+00 to 2+50:

$V = 50(170+65)/2 = 5875$ ft^3

2+50 to 3+00:

$V = 50(65+0)/2 = 1625$ ft^3

Then simply add the volumes and convert to cubic yards

3875+8125+5875+1625 = 19500 ft^3 (0.037037 yd^3/ft^3) = 722.22 yd^3. The Answer is **(A)**

Solution M2

A basic knowledge of formwork is needed for this problem. Since the footing has already been poured, formwork will only be needed for all of the faces of the wall except for the top of the wall. The solution is to add up the surface area of the wall faces.

Rear face walls

13.5'(8.0') + 10.0'(8.0') = 188 ft^2

Front face walls

(10.0' - 1.5')8.0' + (13.5'-1.5')8.0' = 164 ft^2

Outer walls

(1.5'(8.0'))(2 walls) = 24 ft^2
Add up the surface areas = 188 + 164 + 24 = 376 ft^2. The answer is **(C)**

Solution M3

The Critical path of a schedule is the sequence of tasks that determine the minimum amount of time needed to complete the project. Any task on a critical path whose duration is changed will affect the overall schedule of the project.

To solve first develop the task flow chart as shown below. To do this go to each task individually and ask the question "which tasks have this one as a predecessor?". Then draw and connect those tasks with arrows to create a path. Do the same with each task individually until you reach the final task. Then add up the durations for each potential path and the critical path is the largest duration.

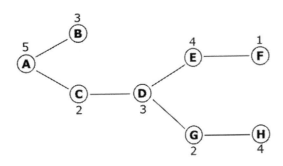

The critical path is tasks A-C-D-G-H with a duration of 16 months. The answer is **(C)**

Solution M4

The initial cost is already at present worth so there is no adjustment of value. The maintenance costs each provide a future value which needs to be converted into a present value

What we have:

n=10 and 20
F= 10,000 and 15,000
i= 3%

What we need: P

Equation needed: $P = F(1+i)^{-n}$

1st maintenance cost: $10,000(1+0.03)^{-10} = 7441$

2nd maintenance cost: $15,000(1+0.03)^{-20} = 8305$ Then add together all present worth costs = 100,000 + 7441 + 8305 = 115746. The answer is **(D)**

Solution M5

First determine the plasticity index: PI = LL − PL = 55-23 = 32

The 50% finer than No. 4 sieve indicates a coarse-grained sandy soil.

Using the word "fines" in a problem is another way to define the No. 200 sieve. Since 20% is greater than 12%, this narrows it down to SM or SC.

A PI over 7 then indicates SC. The Answer is **(A)**

Solution M6

The maximum moment of a simply supported beam is at midspan. Add together the moment from the distributed load and the point load. This can be done by statics or by using the predetermined equations. It's always best to save time by using design aids.

Moment from distributed load = $wL^2/8 = 2(20^2)/8 = 100$ k-ft
Moment from point load = $PL/4 = 3(20)/4 = 15$ k-ft
Total moment = 100 + 15 = 115 k-ft. The Answer is **(B)**

Solution M7

The end conditions dictate the effective length of a compression member. A pinned-pinned condition is 1.0. Be aware in questions like this of if the desired value is the theoretical (ideal) number or the design number.

The answer is **(A)**

Solution M8

The first step is to determine the magnitude and direction of the support reactions

Sum moments about the first support A = 5(5') + 10(10') + 5(15') – B(20') Solve for the right support, B:

B=10 Kips, A = (5 + 10 + 5) – 10 = 10 kips

Then you are able to layout the shear diagram using the following rules:

- To find the magnitude at any point, take a free body diagram from that point to the left-most support and add up the reactions.
- Reactions and loads pointing up are positive, those pointing down are negative
- The shear diagram is flat between concentrated loads
- The shear diagram is sloping along distributed loads

Begin constructing at the left of the beam and work right. The first point is equal to the reaction at A and is upward so the first point is at 10 kips. There is no change until at 5' along the length there is a loading of 5 kips. Therefore, the graph drops to 10 – 5 = 5. Follow this trend to develop the graph.

The answer is **(A)**

Solution M9

First determine the reactions at the left support, A:

Sum moments about point the right support, B = A(10') – 2 k/ft(10')(5') - 5'(2), A = 11 Kips

Since there is no easy equation for the moment at this point on this beam, use the following rule about moment diagrams:

- The magnitude of the moment at any point is equal to the area under the shear diagram curve up to this point

Begin by constructing the shear diagram. We can save time by stopping at point C since we don't need to know any additional information. The graph begins with the reaction at A of 11 kips. You can then find the shear at point C by subtracting the magnitude of the load up to this point.

Shear at C = 11 – 2 k/ft(4') = 3 kips. Therefore, the graph now looks like the figure below. Divide it into sections and find the area:

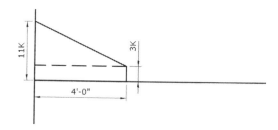

Area 1 = 3 k(4') = 12 kip-ft

Area 2 = ½ (11k - 3k)(4') = 16 kip-ft   Total Moment = 12 + 16 = 28 K-ft The answer is **(C)**

Solution M10

6" diameter concrete cylinders are a common method of testing concrete compressive strengths. The stress from an axial load is determined by P/A.

Area of each cylinder = $\pi r^2 = \pi(3)^2 = 28.27$ in^2

Cylinder 1 stress = 102000/28.27 = 3607 psi
Cylinder 2 stress = 111000/28.27 = 3925 psi
Cylinder 3 stress = 100500/28.27 = 3554 psi

Average of the cylinders = (3607 + 3925 + 3554)/3 = 3695 which is most nearly 3700 psi. The answer is **(D)**

Solution M11

You can find the length of the curve by subtracting stations L = 2+25 – 1+50 = 75'
Convert the interior angle to a decimal 15'/60 = 0.25 Therefore I = 10.25°
The radius can be found by L = (2πrI)/360° = 75' = 2πr(10.25)/360 solve for r = 419.2'

The answer is **(B)**

Solution M12

Cross sections have ideal dimensions to most efficiently carry water. For rectangles, this is achieved by setting the depth equal to the width/2. Options A and B are rectangles but neither satisfies this requirement. For trapezoids, the ideal section is when the depth is equal to two times the Hydraulic Radius. If we take option C and calculate:

Depth = 4cos(30) = 3.5'

R = Area/Wetted Perimeter

$$A = \frac{a+b}{2}h = \frac{4 + (4 + 4\cos(60)\,2)}{2}(3.5) = 21$$

R = 21/(4+4+4) = 1.75

2R = 1.75(2) = 3.5 = d

Therefore, C is an efficient section

The answer is **(C)**

Solution M13

The elevation at the midpoint is at a station L/2 away from the PVC = 800/2 = 400' or 4 stations. We can determine the elevation at the PVI by the following:

Elev$_{PVI}$ = Elev$_{PVC}$ + G$_1$(L/2) = 260 + -(2.5)4 = 250'

Then you can add the distance E to get the point on the curve:

$$E = \frac{|G_2 - G_1|L}{8} = \frac{5(8)}{8} = 5'$$

Elev of Curve = 250 + 5 = 255.0'

The answer is **(D)**

Solution M14

Use the equation for the middle ordinate of a horizontal cure

M=R(1-cos(28.65S/R)) = 700(1-cos(28.65(220)/700)) = 8.63'

The answer is **(A)**

Solution M15

Working backwards from compaction, apply adjustment factors at each stage and carry over the volume:

To achieve 100 yards of 90% compaction you will need $100/0.9 = 111.11$ yd^3

5% is lost so you will need an additional 5% to account for this $111.11(1.05)= 116.65$ yd^3

A swell factor is the volume of the loose excavation material to the in-place excavation material To find the in-place material, divide by the swell factor $116.65/1.07= 109.03$ yd^3

The answer is **(D)**

Solution M16

To solve mix design problems follow the units.

Determine the weight needed in water $W_{water} = W_{cement}(\text{w/c ratio})=500(0.65)=325$ lb/yd^3

The specific weight of water can always be taken as 62.4 lb/ft^3. Use the specific weight to convert a weight to a volume

$V_{water} = W_{water}/\gamma_{water}= (325 \text{ lb/yd}^3)/(62.4 \text{ lb/ft}^3) = 5.21$ ft^3/yd^3 The answer is **(A)**

Solution M17

There are 3 conditions which will simplify the needed equations.

1. Friction is neglected between the wall and soil. This make the angle of external friction $\partial=0$
2. The backfill is horizontal. This makes the slope of backfill $\beta=0$
3. The wall face is vertical. This makes $\theta=0$

The equations needed are now

The active earth pressure coefficient $k_a=\tan^2(45°-\phi/2)$
The total active resultant $R_a= \frac{1}{2} k_a\gamma H^2$

$k_a=\tan^2(45-25/2) = 0.406$
$R_a= \frac{1}{2} (0.406)(115pcf)(15.5)^2= 5606$ lbs

The force is applied at H/3=15.5/3 = 5.167'

The moment is 5.606k(5.167') = 28.9 k-ft The answer is **(C)**

Solution M18

The void ratio is the volume of the voids divided by the volume of the solids.

If a sample is saturated it can be assumed that there are no air voids and thus the total volume of the voids is equal to the volume of the water in that state.

When a sample is dried, there is no more water left in the sample and we have simply the weight of the soil.

Therefore, we are left with the following values

$W_W = 50 - 42 = 8$ lbs

$W_S = 42$ lbs

Since the density of water is known we can then convert the weight to volume

$V_W = 8/62.4 = 0.128$

And since the sample was saturated, all voids are filled with water and $V_w = V_v$

We then need to determine the volume of soil from the weight. Since we have the Specific Gravity (SG), we can determine the density by multiplying by the density of water

$\gamma = 2.4(62.4) = 150 \text{ lb/ft}^3$

$V_s = 42/150 = 0.28$

Void Ratio = 0.128/0.28 = 0.46 The answer is **(D)**

Solution M19

The total distance is a sum of two components. The first is before breaking and the second is after breaking. This is represented in the following equation:

$S_{stopping} = vt_p + S_b$

The first component assumes the velocity is constant during perception reaction and is simply (100 ft/s)(2 seconds) = 200 ft

The stopping distance when breaking occurs is the following:

$S_b = v^2_{mph}/(30(f+G))$

Since the slope is on an incline G is positive. If it was on a decline, it would be negative.

Convert the velocity from ft/s to mi/hour

100ft/sec(3600 sec/hr)/5280 ft/mi = 68.2 mi/hr

$S_b = (68.2^2)/(30(0.30+0.02)) = 484.5 \text{ ft}$

$S_{stopping} = 200 + 484.5 = 684.5 \text{ ft}$ The answer is **(B)**

Solution M20

The superelevation transition distance is the distance between the beginning of the transition to superelevation to the point of being fully superelevated which is the combination of the tangent (also known as crown) runout and the superelevation runoff. The answer is **(A)**

Solution M21

Use the Darcy equation:

$h_f = (fLv^2)/(2Dg)$

Use 32.2 ft/sec^2 as g and then plug in and solve for v

$3.0 = (0.02)(500)v^2/2(2)(32.2)$

v = 6.2 ft/s The answer is **(C)**

Solution M22

The appropriate equation in open channel flow is the Chezy-Manning equation:

$Q = (1.49/n)AR^{2/3}S^{1/2}$

First calculate the hydraulic radius which is the area of water divided by the wetted perimeter which is the perimeter of the sides of the channel which are in contact with water.

R = (6*3)/(6+3+3) = 1.5

$Q = (1.49/0.015)(6*3)(1.5)^{2/3}0.003^{1/2} = 129.3$ The answer is **(A)**

Solution M23

For the given information, the Hazen-Williams equation is appropriate:

$h_f = 10.44LQ^{1.85}/C^{1.85}d^{4.87}$

$40=10.44(100)Q^{1.85}/(140)^{1.85}(3)^{4.87}$

Solve for Q = 432 gpm. The answer is **(D)**

Solution M24

The time base is the amount of time that the flow exceeds the base flow. The answer is **(C)**

Solution M25

The flow rate of the pipe can be used to determine the the area required to limit the flow velocity

The flow rate of the pipe can be determined by the conservation of flow principle

$Q_1 + Q_2 = Q_3$

The flow of the drainage areas can then be determined by the Rational method:

$Q = ACi$ Therefore $Q_3 = 15(0.18)(1.5) + (10)(0.22)(1.5) = 7.35$ cfs

Use the calculated velocity to find the required area $Q/V = A = 7.35/0.26 = 28.28$ ft^2

The answer is **(A)**

Solution M26

When determining how many 0-Force members a truss has, analyze each joint individually as a free body diagram and follow these guidelines:

4. In a joint with 2 members and no external forces or supports, both members are 0-force
5. In a joint with 2 members and external forces, If the force is parallel to one member and perpendicular to the other, then the member perpendicular to the force is a 0-force member.
6. In a joint with 3 members and no external forces, if 2 members are parallel then the other is a 0-force member

Therefore, each joint can be analyzed as follows:

Joint A and H –These joints have 2 members and an external force so refer to guideline 2. Neither member parallel to the force therefore all members are non-zero

Joint C, D, and G – These joints have 3 members and no external force so refer to Guideline 3. 2 members are parallel therefore the other is a 0-Force member. Therefore, Members C-B, D-E, and F-G are 0-Force members

Joints B, E, and F – These joints do not meet any of the guidelines. Therefore, all members are non-zero

The answer is **(C)**

133

Solution M27

This question is provided simply to illustrate that the analysis of the external forces of a truss can be determined in the same way as beams. There may be a tendency to perform unnecessary computations. Simply apply basic statics:

Sum the Forces about H = 0 = -40A + 5(30) + 5(10), Solve for A = 5 Kips

The Answer is **(B)**

Solution M28

First determine the reaction at A. See solution M27.

Determine the angle of triangle A-B-C $Tan^{-1}(8/10) = 38.66$

Using the method of joints, analyze joint A using a free body diagram

Sum the forces in the Y-Direction 5 = ABsin(38.66), AB = 8.00

Sum the forces in the X-Direction AC = ABcos(38.66) = 8cos(38.66) = 6.24

Analyze Joint C

Sum the Forces in the X-Direction AC = CE = 6.24

The answer is **(C)**

Solution M29

Rebar corrosion is most often the result of a chemical reaction due to the intrusion of chlorides into existing concrete. The answer is **(A)**

Solution M30

The water to cement ratio is inversely proportional to the concrete strength and therefore has a direct impact. The answer is **(A)**

Solution M31

Use the Bernoulli equation for the conservation of energy

$E_t = E_{pr} + E_v + E_p = p + v^2/2g + z$

At the water surface of the reservoir, the total energy is the potential energy only and $E_t = 0 + 0 + 50 = 50$

Due to the conservation of energy the total energy at the water surface is equal to the total energy at the exit of the pipe. Therefore the total energy at the exit
$E_t = 50 = 0 + v^2/2(32.2) + 0$, solve for v = 56.75 ft/s. The answer is **(B)**

Solution M32

The answer is **(D).**

The other 3 assumptions are essential to the Bernoulli equation.

Solution M33

Use the conservation of flow principle. $Q_1 + Q_2 = Q_3$ and therefore $A_1V_1 + A_2V_2 = Q_3$

$Q_3 = 3.0(1.2) + (4.0)0.8 = 6.8$ ft^3/s The answer is **(D)**

Solution M34

The effective stress is the density times the height of each level. However, if the water table is present, the density is reduced by that of water:

Effective Stress = 130(20) + 100(10) + (100-62.4)(10) = 2600 + 1000 + 376 = 3976 psf

The answer is **(C)**

Solution M35

The Maximum bending stress is determined at the location of maximum moment.

First determine the reactions at the supports

Sum moments about B = 0 = -15A + 10(10) + 10(5), A = 10

Due to symmetry A = B = 10

The Max moment for a beam with 2 symmetrical point loads = PX where X is the distance to the first point load = 5'(10 kips) = 50 Kips-ft

Then apply the equation for bending stress Mc/I

Moment of inertia of a rectangle $(1/12)bh^3 = (1/12)(4)(6)^3 = 72$ in^4

The bending stress = $(50$ k-ft$)(12$ in/ft$)(3$ in$)/72$ in$^4 = 25$ ksi

The answer is **(A)**

Solution M36

$P/A = 0.1$ ksi $= 9/(12w)$, w = 7.5"
B can be determined by finding the additional horizontal distance from the critical section for shear to the base.

$$\frac{1}{8} = \frac{x}{6} \quad x = 0.75"$$

b = 7.5" + 0.75" = 8.25"

The answer is **(C)**

Solution M37

The determination of which roadside safety barrier system is determined by the distance from the edge of road. Barriers can be flexible, semi rigid, or rigid. A concrete barrier is rigid and is for locations where zero deflection is needed. A metal beam rail is semi-rigid and is used for objects or drop-offs within a short distance of the road. Cable systems are flexible and have the most deflection and are used for objects or drop-offs further from the road.

The answer is **(C)**

Solution M38

The peak hourly traffic volume is the hour long timeframe in which the most cars are observed. First calculate the total volume for each hour long time frame:

8:00-9:00 = 500 + 560 + 650 + 625 = 2335
8:15-9:15 = 560 + 650 + 625 + 630 = 2465
8:30-9:30 = 650 + 625 + 630 + 600 = 2505
8:45-9:45 = 625 + 630 + 600 + 540 = 2395
9:00-10:00 = 630 + 600 + 540 + 460 = 2230

Then determine the largest volume = 2505 The answer is **(C)**

Solution S39

The equation for deflection at the end of a cantilever beam with a point load is:

$Pl^3/3EI$

First determine the moment of inertia:

$I = (1/12)(6)12^3 = 864 \text{ in}^3$

$\Delta = 1(10(12))^3/(3(3605)(864)) = 0.18''$, The answer is **(B)**

Solution M40

The specific gravity is the density of the soil over the density of water.

The volume of the voids can be determined by subtracting the soil volume from the total

$V_V = V_T - V_S = 2.0 - 1.5 = 0.5 \text{ ft}^3$

The degree of saturation can be used to determine the volume of water

$V_W = V_V(S) = 0.5(0.75) = 0.375 \text{ ft}^3$ Convert volume to weight $W_W = 0.375(62.4) = 23.4 \text{ lb}$

Then determine the weight of soil thought the moisture content

$W_S = W_W/w = 23.4/0.1 = 234 \text{ lbs}$, the density is then $234/1.5 = 156$

The SG = $156/62.4 = 2.5$ The answer is **(B)**

Solution M41

First, the % passing of fines is another term for the % passing the no. 200 sieve. Using the AASHTO Classification chart, this narrows the category down to the A-2 class. Then determine the Plasticity Index:

Plasticity Index = Liquid Limit − Plastic Limit = 36 − 24 = 12

Then using the LL=36 and PI=12, the appropriate Class is A-2-6.

The Answer is **(C)**

Solution M42

The Charpy V-Notch is a test for the toughness of steel.

The answer is **(D)**

Solution M43

First categorize the loads as dead or live loads:

Dead: Floor Slab, Flooring, Utilities. These are loads which are permanent to the structure.

Live: Pedestrian, Furniture. These are loads which will or may move.

Then calculate the load:

1.2(100 + 10 + 2) + 1.6(80 + 40) = 326.4 psf

The answer is **(C)**

Solution M44

This beam is fixed on one end and supported on the other with a uniform distributed load. Use the beam chart to find the appropriate equation:

$$\Delta = \frac{wL^4}{185EI}$$

First find the moment of inertia:

I = 1/12bh³ = 1/12(6)(12)³ = 864 in⁴

Then plug into the equation:

$$\Delta = \frac{(2)(10*12)^4}{185(3605)(864)} = 0.72"$$

The answer is **(D)**

Solution M45

The work which takes place within the adjacent property is temporary in nature and the current owner will maintain rights of the property and therefore a temporary easement would be required.

The answer is **(A)**

Solution M46

Determine the quantities for each item in the appropriate unit.

Concrete Wall = (10(1) + 8(2.5))20 = 600 cu. ft.

Backfill = ((4)10 + ½(2)(10))20 = 1000 cu. ft./27 = 37.04 cu. yd.

Drain = 20'

Then use the unit prices to find the total cost:

600(30) + 37.04(100) + 20(20) = 22104.

The answer is **(B)**

Solution M47

The answer is **(B)**

Solution M48

First convert the load in tons to lbs. 75 tons (2000 lbs/ton) = 150,000 lbs

Then using the chart for a 120' boom, the most appropriate answer is 30'.

The answer is **(B)**

Solution M49

Due to the desired accelerated time of construction, cast in place can be eliminated due to the need for curing times. Then the best choice is dictated by the span length. 80' is much longer than the capacity for precast concrete and post tensioned concrete is not economical at this span. Therefore the best choice is Prestressed-precast concrete.

The answer is **(C)**

Solution M50

The length L can be determined by setting the reaction at the temporary support to the maximum allowable of 5 kips. Sum moments about the end of the beam:

0 = 0.288 k/ft(30')(30'/2) – L(5 K) Solve for L = 25.92'

The answer is **(C)**

Solution M51

The need for Temporary earth retaining is based on the soil type and the depth of cut. The depth of cut without TERS is limited to less than 5 feet. Also, stable rock is a type of soil that is considered sufficiently safe without support. Therefore the answer is A

The answer is **(A)**

Solution M52

$$F = \frac{N_o c}{\gamma_{eff} H}$$

$$\gamma_{eff} = 125 - 62.4 = 62.6\ pcf$$

$$1.5 = \frac{6.2(300)}{(62.6)H} \qquad H = 19.8'$$

X = 19.8/1.5 = 13.2

Available Right of way = 100 − 2(13.2) = 73.6'

The answer is **(C)**

Solution M53

$$h = \Sigma k \frac{v^2}{2g}$$

$$K_{90} = 0.9$$

$$K_c = 0.5\left[1 - \left(\frac{D_1}{D_2}\right)^2\right] = 0.5\left[1 - \left(\frac{8}{12}\right)^2\right] = 0.277$$

$$A = \pi r^2 = \pi 0.33^2 = 0.4\ ft^2$$
$$V = Q/A = 2.0/0.4 = 5\ ft/s$$

$$h = (0.9 + 0.277)\frac{5^2}{2(32.2)} = 0.46'$$

The answer is **(B)**

Solution M54

The Euler critical stress is the following equation:

$$F_{Cr} = \frac{\pi^2 E}{\left(\frac{KL}{r}\right)^2}$$

Where r = Radius of Gyration = $\sqrt{\frac{I}{A}}$

I = 1/12(36)(36)^3 = 140000 in^4

$$r = \sqrt{\frac{140000}{36(36)}} = 10.4 \; in$$

The K can be determined by the end conditions, for fixed and pinned the design value is 0.8 from the chart.

$$F_{Cr} = \frac{\pi^2 (3605)}{\left(\frac{0.8((25)(12))}{10.4}\right)^2} = 66.8 \; ksi$$

The answer is **(D)**

Solution M55

Using the USCS Classification chart, the description for the group symbols is as follows

GW – Well graded gravels
GP – Gap-graded or uniform gravels
SP – Gap-graded or uniform sands
CH – Highly plastic clays

Primary consolidation is a gradual consolidation as water leaves the voids over time. This is mostly an issue in clay type soils. Therefore CH can be expected to see the most primary consolidation.

The answer is **(D)**

Solution M56

A 100-yr storm event is a storm of a design intensity which is only anticipated to occur once every 100 years.

The answer is **(C)**

Solution M57

The flow in needs to equal the flow out of a system. However, since the flow out is limited, this will produce a backup of water which will begin to fill the pond. First determine the flow rate of the outlet:

$$Q = VA = (0.25)(\pi r^2) = (0.25)(\pi(1)^2) = 0.785 \text{ ft}^3/\text{s}$$

Then the excess can be determined by subtracting the flow out from the flow in = 0.85 − 0.785 = 0.065 ft³/s

Then determine the size of the pond by determining how much volume of water will accumulate during the storm:

$$V = 0.065(6 \text{ hr})(60 \text{ min/hr})(60 \text{ s/min}) = 1404 \text{ cu. ft.} = 52 \text{ cu. yd.}$$

The answer is **(B)**

Solution M58

The N-value is the number of blows required to drive the sampler 12" after the initial 6". Therefore the numbers to average are 26, 38, and 60. The average is 41.33.

The answer is **(D)**

Solution M59

The permeability of soil is determined by Darcy's Law:

$$Q = KiA$$

First the flow rate can be determined from the sample by dividing the volume by the time:

$$Q = 1/10 = 0.1 \text{ in}^3/\text{s}$$

Then determine the cross-sectional area of the sample:

$$A = \pi r^2 = \pi(3)^2 = 28.27 \text{ in}^2$$

The hydraulic gradient, I, is determined by the difference in pressure head divided by the length;

i = 3/12 = 0.25

Then solve for K = Q/iA = 0.1/((0.25)(28.27)) = 0.014 in/s

The answer is **(B)**

Solution M60

For different aspects of construction, the required level of accuracy varies. For the construction of bridges this is 0.01. Therefore the measurement should be recorded as 158.57'.

The answer is **(B)**

Solution M61

When a project is complete, areas which have been excavated should be reestablished with seeding to promote growth. The other methods are temporary measures during construction.

The answer is **(C)**

Solution M62

The compressive stress for a concentric point load is P/A. However since this is eccentric, an additional stress due to the moment must be applied. Therefore:

$$f_c = \frac{P}{A} + \frac{MC}{I}$$

I = 1/12(14)(14)³ = 3201 in⁴

$$f_c = \frac{5}{14(14)} + \frac{5(6)(7)}{3201} = 0.091 \; ksi$$

The answer is **(A)**

Solution M63

The choice of an appropriate unit of measure needs to be a representation of what varies from project to project as well as what can be easily measured. In this case the box culverts have a consistent area and therefore a measure of area would not represent the work appropriately. The unit each would be difficult to estimate since box culvert sections can vary based on the manufacturer's capabilities. This would be difficult to approximate the quantity. Lump sum is used for general tasks which may be difficult to measure. What does vary, is known, and can be easily measured is the length of the box culverts. Therefore the appropriate unit is Linear ft.

The answer is (A)

Solution M64

The critical section for moment on a spread footing is at the face of the column. For a point load only the equation for the moment due to the uniform bearing pressure is:

M_u = q_uLl^2/2 = (50/(10(10)))(10)(4)^2/2 = 40 K-ft The answer is (D)

Solution M65

The time of concentration is a product of three components. However in this example we are only looking for sheet flow time. Therefore:

$$t_c = t_{sheet} = \frac{0.007(nL_o)^{0.8}}{\sqrt{P_2}S_{deciaml}^{0.4}} = \frac{0.007((0.15)(250))^{0.8}}{\sqrt{2.2}(0.05)^{0.4}} = 0.284 \ hrs$$

The answer is (A)

Solution M66

The Peak Hour Factor is ratio of the hour interval with the greatest volume to four times the largest 15 min. interval volume. First determine the hour interval with the greatest volume

7:00-8:00	1900+2200+2350+2000=8450
7:15-8:15	2200+2350+2000+1700=8250
7:30-8:30	2350+2000+1700+1200=7250

Therefore the largest hourly volume is 8450 in interval 7:00-8:00. The 15-min. interval with the greatest volume in this interval is 7:30-7:45 with a volume of 2350.

$$PHF = \frac{V_{vph}}{4V_{15\ min,peak}} = \frac{8450}{4(2350)} = 0.8989$$

The answer is **(C)**

Solution M67

Use the NRCS method

S = 1000/CN − 10 = 1000/79 − 10 = 2.658

$$Q = \frac{(3.5 - 0.2(2.658))^2}{3.5 + 0.8(2.658)} = 1.56\ in$$

The answer is **(A)**

Solution M68

Use the Terzaghi Bearing Capacity equation:

$$q_{ult} = \frac{1}{2}\gamma B N_\gamma S_\gamma + c N_c S_c + (p_q + \gamma D_f)N_q$$

However, since the soil is cohesionless and there is no surcharge load, the equation becomes as below. Also since the footing is a strip type, the shape factors are 1.0 and the equation becomes:

$$q_{ult} = \frac{1}{2}\gamma B N_\gamma + (\gamma D_f)N_q = \frac{1}{2}(0.125)(8)(2.5) + (0.125)(2.5)(4.4) = 2.625\ ksf$$

The ultimate pressure then needs to be adjusted for overburden:

$q_{net} = q_{ult} - \gamma D_f = 2.625 - 0.125(2.5) = 2.3125$ ksf

Finally calculate the allowable from the net pressure:

$q_a = q_{net}/FS = 2.3125/2.5 = 0.925$ ksf

The answer is **(A)**

Solution M69

A mat foundation is a large scale pour which would require a slower curing process to limit cracking and reduce heat release. The types of cement which are appropriate for this are type II and IV. Type II is more suitable for protection against sulfate attack and therefore the choice is Type IV.

The answer is **(D)**

Solution M70

V = Area under Hydrograph

Lag Time = Time from greatest rainfall to peak discharge = Time from hour 2 to hour 6

1 hr (60)(60) = 3600 seconds

3600(1/2(0.5) +1.4 +2.5 + 6.6 + ½(8.6)) = 5480 cf

The answer is **(B)**

Solution M71

To prevent cracking, the modulus of rupture needs to be greater than the applied stress. First determine the maximum moment in the wall panel. Since it can be approximated as a simply supported beam with uniform load, the max moment is:

$$M_u = \frac{wl^2}{8} = \frac{(0.8)(18)^2}{8} = 32.4k - ft = 388.8k - in$$

Calculate the moment of inertia of the resisting section:

I = 1/12((8)12)(8)³ = 4096 in⁴

Then determine the stress due to bending:

$$f_b = \frac{Mc}{I} = \frac{(388.8)(4)}{4096} = 0.380 \ ksi = 380 \ psi$$

Then equate the stress to the equation for the modulus of rupture to find the minimum strength:

$$7.5\sqrt{f'_c} = 380$$

f'_c = 2567 psi

The answer is **(C)**

Solution M72

In this example the present worth of each project is known and the future value at different times is needed. The equation to turn a present value to a future value is:

$$F = P(1 + i)^n$$

Use this equation for each project:

$$F_1 = 100000(1 + 0.03)^3 = 109273$$

$$F_2 = 175000(1 + 0.03)^4 = 196964$$

$$F_3 = 225000(1 + 0.03)^5 = 260837$$

Then add up the total cost = 567073.

The answer is **(D)**

Solution M73

For flexure the important factor is the moment of inertia. Unlike the other members, the WT shape does not have a bottom flange portion and has a poor distribution of area. Therefore this would be the least effective in flexure.

The answer is **(B)**

Solution M74

The strength of the steel has no effect on the steels ability to resist corrosion. Therefore High Strength Steel does not assist in corrosion protection.

The answer is **(C)**

Solution M75

The sandy soil creates a stable bearing condition which indicates that a deep foundation element is not necessary. Also note that the bridge does not cross a waterway. Often for scour prevention measure, bridges which cross water are required to be on deep foundation elements. Therefore a spread footing should be sufficient.

The answer is **(A)**

Solution M76

Because of limited resources, the crews have to be assigned work. Crews can work independently or simultaneously.

A = 4 days per crew/2 crews = 2 days

B + C = 12/2 = 6 days

D + E = 18/2 = 9 days

F = 10/2 = 5 days

Total = 2 + 6 + 9 + 5 = 22 days

Total cost of labor = 22(8 hrs/day)(50 $/hr)(4 workers) = $35200

The answer is **(C)**

Solution M77

For the existing schedule the breakdown of paths and durations are as follows:

Path	Total Duration
A-B-C-E	22
A-B-C-F	21
A-B-D-F	19
A-B-D-G	21

Therefore the critical path is A-B-C-E. The durations will then change and the new critical path needs to be determined. The revised schedule is shown below:

Path	Total Duration
A-B-C-E	23
A-B-C-F	25
A-B-D-F	26
A-B-D-G	25

The critical path now becomes A-B-D-F with a duration of 26.

The answer is **(D)**

Solution M78

In the design of slabs, the analysis is performed by taking a 1' wide strip of the slab as a beam. Therefore this can be analyzed by taking a 1' wide beam which is fixed at both ends with a distributed load. The maximum moment for this loading and support condition is:

$$M = \frac{wl^2}{12} = \frac{(0.5\ ksf)(1')(20)^2}{12} = 16.67\ k - ft$$

The answer is **(D)**

Solution M79

Offset distance to the Baseline of road is indicated as 16'. This point has an elevation of 144'

Distance from centerline of road to face of curb = 16' – 4' =12'

Calculate drop in elevation from the cross slope 144' – 0.02(12) = 143.76'

Add height of curb 143.76 + 10/12 = 144.60'

The answer is **(C)**

Solution M80

Find the yield strength by identifying the point at which the load decreases and there is an appreciable change in elongation compared to the increase in load in this case it can be identified at load 2656 lbs.

Strength = P/A = $2656/(\pi r^2)$ = $2656/(\pi(0.125)^2)$ = 54.1 ksi

The answer is **(C)**

Solution T1

To determine level of service we must first determine the density and then use the appropriate chart to match the level of service to the calculated density. The density is:

$$D = \frac{v_p}{S}$$

Let's first calculate the peak flow rate:

$$v_p = \frac{V}{(PHF)N f_{HV}}$$

The heavy vehicle factor is the following:

$$f_{HV} = \frac{1}{1 + P_T(E_T - 1)}$$

The percentage is given as 5%. The passenger car equivalent can be determined from exhibit 12-25. Since the stretch is < 1 mile and the grade is level. E_T = 2.0.

$$f_{HV} = \frac{1}{1 + 0.05(2.0 - 1)} = 0.95$$

$$v_p = \frac{1400}{(0.93)3(0.95)} = 528 \; pcphpl$$

The speed can be determined from the field observations

$$S_{avg} = \frac{N_t L}{\Sigma t}$$

First sum the times t = 34 + 43 + 44 + 48 + 52 + 31 + 36 + 37 + 45 + 48 = 418 s

$$S_{avg} = \frac{10(0.75)}{418} = 0.018 \; mps = 64.6 \; mph$$

Now calculate the density (note the speed does not need to be adjusted due to the geometry of the roadway):

$$D = \frac{528}{64.6} = 8.17 \; pcpmpl$$

From HCM Exhibit 12-15, the LOS = A

The answer is **(A)**

Solution T2

The acceptable clear zone can be determined from the slope, design speed, and AADT from Table 3-1 in the RSDG. Therefore, you can determine the following values:

Interior of curve = 14-16', use 14'
Exterior of curve = 16-20' use 16'

Note we took the lower values of the range since we are looking for the minimum. However, the outside of the curve must be adjusted. From table 3-2, the adjustment factor is 1.2. Therefore:

X = 14' + 1.2(16') + 24' = 57.2'

The answer is **(D)**

Solution T3

The effective width is the total width of walkway minus effective width reductions. In this case we have a bench in the walkway. From exhibit 24-9, we can determine that benches have a preempted width of 5'. As described in the notes, widths are from curbs to the edge of objects. However, the note also says an additional shy distance must be added to widths of 1.0'-1.5'. Since we are using the maximum of the ranges, use 1.5'. The zone of influence of the bench and the store fronts are shown below:

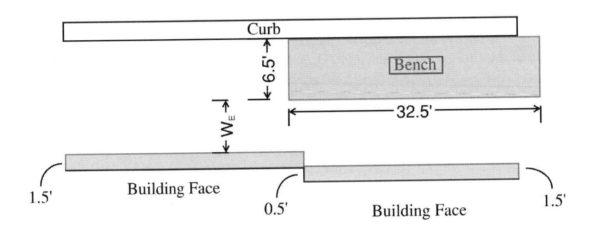

The length of the influence from the bench can be taken as 5 times the width so in this case 5(6.5) = 32.5'. This causes the bench to overlap the closer building face and the effective width is as shown and equals:

$W_E = 12 - 6.5 - 1.5 = 4'$

The answer is **(A)**

Solution T4

Trip generation data is often approximated by using straight line approximation. We can use the following equation:

$$T = y + mX$$

First we can determine the slope of the line. The data points given are all on the same line so comparison of any two points works:

$$m = \frac{900 - 770}{390 - 325} = 2$$

To determine the y-intercept, plug in any point:

$$900 = y + 2(390); y = 120$$

Therefore, the trip generation equation is:

$$T = 120 + 2X$$

Plug in 450:

$$T = 120 + 2(450) = 1020 \ trips$$

The answer is **(D)**

Solution T5

The equation for the saturation flow rate is the following:

$$s = s_o f_w f_{HVg} f_p f_{bb} f_a f_{LU} f_{LT} f_{RT} f_{Lfb} f_{Rfb} f_{wz} f_{ms} f_{sp}$$

The equation is lengthy and includes many factors so you may feel compelled to start looking individually at each factor. However, we are not given much information. Since the base and the calculated rate are provided, we can determine the product of all the factors:

$$s = s_o f_{product}; \ f_{product} = \frac{2500}{1900} = 1.316$$

The only changes made are to the lane width and the grade. We can now calculate the new lane width and grade adjustment factors and compare them to the old:

$f_{w,old} = 0.96\ f_{w,New} = 1.00\ from\ exhibit\ 19-20$

Then calculate the increase from old to new:

$$f_{w,increase} = \frac{1.0}{0.96} = 1.042$$

Now you can calculate the modified saturation flow rate = 2500(1.042) = 2605 pcphpl

The answer is **(B)**

Solution T6

The predicted crash frequency is the following equation:

$$N_{Predicted} = N_{SPF}C(CMF)$$

The Safety Performance Function frequency equation is given so we can calculate:

$$N_{SPF} = (9)e^{-1.2}(1250)^{0.5} = 95.83\ \text{crashes/year}$$

Then we can determine the crash modification factor for each of the crash sites:

$$CMF_1 = \frac{14}{25} = 0.56$$

$$CMF_2 = \frac{12}{21} = 0.57$$

$$CMF_3 = \frac{16}{31} = 0.52$$

Then calculate the predicted frequency;

$$N_{Predicted} = (95.83)1.1(0.56 \times 0.57 \times 0.52) = 17.5, use\ 18\ crashes/yr$$

The answer is **(B)**

Solution T7

To determine the minimum length of curve, we must determine the difference in the grades, A. We can correlate the middle ordinate to A by the following where X is the PVC station:

$$M = \frac{AL}{800} = 3 = \frac{AL}{800}; A = \frac{2400}{L}$$

Then use the appropriate equation from the AASHTO GDHS. Let's first guess L > S:

$$L = \frac{AS^2}{2158} = L = \frac{\left(\frac{2400}{L}\right)155^2}{2158}; L = 163.5'. L > S\ OK$$

Now find the PVC station:

163.5 = (2500 – X); X = 2337 = 23+37

The answer is **(C)**

Solution T8

Since we are given the sum of the pre-maneuver and the maneuver times, we can eliminate A. We can then use the equation:

$$d_{ft} = 1.47v_{mph}t_{t,sec} = 1.47(60)(12.4) = 1093.7\ ft$$

From GDHS Table 3-3 we can match the design speed and sight distance. Since the number is closest to 1125, the avoidance maneuver is D.

The answer is **(C)**

Solution T9

The length of the superelevation runoff is:

$$L = \left(\frac{W N_L e_\%}{\Delta_\%} \right) b_w$$

Let's gather the variables

W = 12
N_L = 3
e = 7
Δ = 0.45 (GDHS Table 3-15)
b_w = 0.67 (GDHS Table 3-16)

$$L = \left(\frac{(12)(3)(7)}{0.45} \right) 0.67 = 375.2'$$

However, the curve does not begin at the beginning of the transition. The PVC begins at 2/3 of the distance and the station of full superelevation is:

Station of Superelevation = 1/3(375.2) + 4500 = 46+25

The answer is **(B)**

Solution T10

The clearance of the bridge is simply the elevation difference between the low point of the bridge and the elevation of the roadway at that station.

First calculate the gradient of the curve A = $(G_2-G_1)/L$ = (-4-2)/8 = -0.75

Determine the beginning of vertical curve (BVC) station:

BVC = 10+00 – 800/2 = 6+00

Determine the elevation of BVC = 155.0 – (.02)400 = 147.0'

Determine the elevation of station 8+00 by using the equation of a parabola (Be careful with signs) where X is the difference in stations from the point of interest to the BVC

Y = $Elev_{BVC}$ + G_1(X) + (A/2)X^2 = 147.0 + 2(2) + (-0.75/2)(2)^2 = 149.5'

Then subtract the road elevation from the low point of the bridge to find the clearance

170.0 – 149.5 = 20.5'.

The answer is **(B)**

Solution T11

The equation for free flow speed of a multilane highway is:

$$FFS = BFFS - f_{LW} - f_{TLC} - f_M - f_A$$

f_{LW} = 0 since the lanes are 12' as per Exhibit 12 – 20
f_{TLC} = Since the highway is divided, the lateral clearances are to be added. Therefore use 6' + 4' = 10' and from Exhibit 12-22, use a 0.4 reduction
f_M = 0 for divided highways
f_A = First determine the access point density per mile after the rehab = 40/10 = 4. Then interpolate from Exhibit 12-24: 4(2.5/10) = 1.0

Then plug into the equation:

$$FFS = 65 - 0 - 0.4 - 0 - 1.0 = 63.6$$

The answer is **(C)**

Solution T12

For a stretch of roadway, the accident rate is determined from the following equation:

$$N_{Segment} = \frac{N_{CR}\,10^8}{(AADT)N_{yr}\left(365\,\frac{days}{yr}\right)L_{mi}} = \frac{(28+31+45)(10^8)}{(1600+1200)(3\,yrs)\left(365\,\frac{days}{yr}\right)(5)} = 678.4$$

The answer is **(D)**

Solution T13

The low point is where the slope of the gradient is zero. Therefore:

X = G₁/A

A = (G₂-G₁)/L = (2-(-3))/6 = 0.833

X = 3/0.833 = 3.6 stations

The low point is at station 7+00 + 3+60 = 10+60.

The answer is **(D)**

Solution T14

Let's breakdown each statement:

A: Max spacing for merging patterns of channelization devices is 1' per each mph of design speed. A is true

B: Since S < 40, L = WS²/60 = (10)(25)²/60 = 104.1'. Then from Table 6C-3 L minimum = L. B is true

C: For an urban street with low speed, the minimum spacing is 100'. C is true

D: Only 3 advance warnings are necessary. D is false

The answer is **(D)**

Solution T15

To determine level of service we must work our way to determining the density and then use the appropriate chart to match the correct level of service. The density is:

$$D = \frac{v_p}{S}$$

D = Density (pc/mi/ln)
v_p = Peak flow rate (pc/hr/ln)
S = Speed (mph)

Since the free flow speed is measured, we can use this as S. We must now calculate the peak flow rate:

$$v_p = \frac{V}{(PHF)Nf_{HV}}$$

Since there are no trucks, the Heavy vehicle factor is also 1.0. However, we still need to calculate the peak hour factor. Since we are given measured intervals, we can use the following:

$$PHF = \frac{V_{vph}}{N_t V_t}$$

In this case we are using 15 min intervals so N_t = 60/15 = 4.

$$PHF = \frac{2050}{4(620)} = 0.83$$

Then calculate the peak flow:

$$v_p = \frac{2050}{(0.83)3(1.0)} = 823\ pcphpl$$

Then the density:

$$D = \frac{823}{60} = 13.7\ pcpmpl$$

Then use exhibit 12-15, the LOS = B

The answer is **(B)**

Solution T16

In this problem we need to understand that each lane has its own curve profile but with the same angle and differing radii. Let's start with the first interior lane and compare the offset of the building to the horizontal sightline offset.

First interior curve:

Calculate the sightline which is also the chord length:

$$S = 2R \sin \frac{I}{2} = 2(4500) \sin \frac{12}{2} = 940.8'$$

Then determine the horizontal offset:

$$H = R\left(1 - \cos \frac{28.65S}{R}\right) = 4500\left(1 - \cos \frac{28.65(940.8)}{4500}\right) = 24'$$

This is the offset of the line of sight for the first interior lane from the center of curve. We need to calculate how far the building is from the first lane:

Offset of building from first lane = 30' − 3 − 12 - 6 = 9'. Since 9' < 24', the sight is obstructed.

Lets now evaluate lane 2, the angle remains but the radius increases the width of one lane R_2 = 4512'. Let's follow the same process for lane 2:

$$S = 2(4512) \sin \frac{12}{2} = 943.3'$$

Then determine the horizontal offset:

$$H = 4512\left(1 - \cos \frac{28.65(943.3)}{4512}\right) = 24.63'$$

The distance to the building from the second lane = 30' − 3 − 6 = 21' < 24'. Lane 2 is obstructed.

Now lane 3, R = 4512 + 12' + 6' = 4530'

$$S = 2(4530) \sin \frac{12}{2} = 947.0'$$

Then determine the horizontal offset:

$$H = 4530\left(1 - \cos\frac{28.65(947.0)}{4530}\right) = 24.72'$$

Distance from building to centerline of 3rd lane = 30' + 6 + 6 = 42' > 24.72'. Lane 3 is clear

Therefore 2 lanes are obstructed.

The answer is **(B)**

Solution T17

The distance to the length of need, X, for a non-flared barrier is the following:

$$X = \frac{L_A - L_2}{\dfrac{L_A}{L_R}}$$

We can determine the Runout from RSDG Table 5-10, L_R = 250'
From Table 5-7 the shy distance = 8'; L_2 = 8' + 2' = 10'

$$X = \frac{16 - 10}{\dfrac{16}{250}} = 93.75'$$

Then determine the station = 2140 – 93.75 = 20+46

The answer is **(B)**

Solution T18

The equation for the skid resistance number can be used to determine the friction factor necessary:

$$SN = 100(f) = \frac{35}{100} = 0.35$$

Then you can determine the design speed from the following superelevation equation:

$$R = \frac{v_{mph}^2}{15(e + f)} = 750 = \frac{v^2}{15(0.05 + 0.35)}; v = 67\ mph$$

The answer is **(C)**

Solution T19

Traffic warrants are covered by the MUTCD. Option C provides a volume ratio which falls below the curve on Figure 4C-1 and therefore does not satisfy the warrant condition.

The answer is **(C)**

Solution T20

The Manual on Uniform Traffic Control Devices provides guidance for when to use or consider certain traffic signs based on engineering judgement. Section 2B.06 provides appropriate applications for stop signs. 2B.06.02a indicates that a stop sign should be considered if the volume on the through street is greater than 6000 vehicles per day. Therefore option B warrants consideration.

The answer is **(B)**

Solution T21

We must compare the current level of service to the future volume based on the growth. We can back figure the future volume by using the lowest density from exhibit 12-15 at which the LOS = F. This is D > 45, so let's say 45.01, We can then calculate the corresponding peak flow rate:

$$D = 45.01 = \frac{v_p}{65}; \ v_p = 2926 \ pcphpl$$

Then determine the corresponding volume:

$$v_p = \frac{V}{(PHF)Nf_{HV}} = 2926 = \frac{V}{0.93(3)(1.0)}; V = 8163 \ pc/yr$$

Then you can use the traffic forecast model to determine the growth rate:

$$V_{Future} = 8163 = (5500)(1 + P)^{15}; P = 0.0267 = 2.67\%$$

The answer is **(A)**

Solution T22

As per the MUTCD Table 3B-1, the minimum passing sight distance for 35 mph is 550 ft. We can then use the appropriate equation for determining the length of the curve. First assume S < L:

$$L = \frac{AS^2}{2800} = \frac{|-3-2|550^2}{2800} = 540.2'; L < S \text{ Therefore use alternate equation}$$

$$L = 2S - \frac{2800}{A} = 2(550) - \frac{2800}{|-3-2|} = 540'; L < S; OK$$

The answer is **(A)**

Solution T23

We need to first determine the required vertical clearance for the provided information. We can use the following equation when S > L:

$$L = 1200 = 2S - \frac{800\left(C - \left(\frac{h_1 + h_2}{2}\right)\right)}{A} = 2(1250) - \frac{800\left(C - \left(\frac{8+2}{2}\right)\right)}{4}; C = 11.5'$$

From the cross section however, we can see that we do not have this amount of clearance across the roadway. Since the slope of the girder and roadway are different, the clearance increases as we move away from the abutment. The distance from the abutment can be found by:

$$11.5' = (90 - 79.3) + (0.06X - 0.02X); X = 20'$$

The answer is **(D)**

Solution T24

First we can determine the deceleration rate needed to stop the vehicle in 15'. First convert mph to ft/s: 45 mph (1.47) = 66.15 ft/s. Then the deceleration rate is:

$$d = \frac{v^2}{2Lx} = \frac{66.15^2}{2(15)(0.85)} = 171 \ ft/s^2$$

The stopping force needed then is:

$$F = \frac{SF(w)d}{g} = \frac{1.2(4000)(171)}{32.2} = 25490 \ lbs$$

Then divide by the weight of barrels: 25490/2000 = 12.75, use 13 barrels

The answer is **(D)**

Solution T25

The Policy for the Geometric Design of Highways and Streets provides figures for different standard design vehicles. The SU-12 is found on figure 2-3. The width needed is from the overhang radius to the inner rear wheel radius. The space needed is:

52.8' – 36.4' = 16.4'

The answer is **(C)**

Solution T26

First determine the number of pedestrians per cycle:

$$N_p = \frac{N_{do}(C - g)}{C} = \frac{\left(\frac{15}{3600}(190)\right)(190 - 40)}{190} = 0.6 \; ped/cycle$$

The length of cross walk then relates to the time to clear by:

$$t = 3.2 + \frac{L}{S_p} + 0.27N_{ped} = 25 = 3.2 + \frac{L}{2} + 0.27(0.6); L = 43.27 \; ft$$

$$Lane \; width = \frac{(43.27 - 3.25)}{4} = 10'$$

The answer is **(A)**

Solution T27

The cycle length is the total of all phases. In this case we must add up the phase time for all three phases. First understand that the effective green is the clear time and the vehicle arrival time plus an extension. This extension however is not included in calculating phase time. Therefore, we can make the phase duration equation the following:

$$D_p = l_1 + (g - 2) + Y + R_c$$

Calculate for each phase:

$$D_{p1} = 1.5 + (10 - 2) + 4 + 2 = 15.5 \; s$$

$$D_{p2} = 1.5 + (30 - 2) + 5 + 2 = 36.5 \; s$$

$$D_{p3} = 1.5 + (20 - 2) + 5 + 2 = 26.5 \; s$$

Cycle length = 15.5 + 36.5 + 26.5 = 78.5 s

The answer is **(D)**

Solution T28

The GDHS chapter 10 provides guidance on the design of interchanges. The type of interchange is often dependent on a large number of variables such as volume, design speed, and rights of way. In this question, the given road types and setting likely can use a less complex interchange type. Therefore, a diamond is most appropriate.

The answer is **(A)**

Solution T29

The sight distance at an intersection is the hypotenuse of a triangle formed by the stopping distance and the offset between the drivers' vision. We are given the offset as 15' + 8' =23'.

Lets then calculate the stopping distance:

$$X = \frac{v_{mph}^2}{30(f + G)} = \frac{55^2}{30(0.35 + 0.02)} = 272.5'$$

Then we can calculate the sight distance which is the hypotenuse of the triangle formed by the stopping distance as the base and the perpendicular distance between the cars as the height = $272.5^2 + 23^2 = 273.5'$

The answer is **(B)**

Solution T30

A compound curve can be analyzed as two separate curves adjoined at a common point. We must first find the station of this common point the PCC which is at the end of the first curve:

$$L_1 = \frac{\Delta \pi R}{180} = \frac{10\pi(2000)}{180} = 350'$$

PCC = 2550 + 350 = 29+00

Since the PCC = PC$_2$, we can use it to determine the angle of the second curve:

$$L_2 = 3245 - 2900 = \frac{\Delta\pi(2500)}{180}; \Delta = 7.9^o$$

The answer is **(C)**

Solution T31

ADA Section 503 covers Passenger loading zones. It consists of a vehicle pull up space which must be a minimum of 96" and an access aisle which must be 60".

96 + 60 = 156"

The answer is **(D)**

Solution T32

The equation for the structural number for flexible pavement with two layers is:

$$SN = D_1 a_1 + D_2 a_2 m_2$$

However, since we are given the individual strength numbers, we can break down the equation as follows:

$$SN_1 = D_1 a_1 = 1.8 = D_1(0.42); \; D_1 = 4.28"$$

$$D_2 = \frac{SN_2 - SN_1}{a_2 m_2} = \frac{4.6 - 1.8}{(0.11)(0.9)} = 28.28'$$

The ratio is then:

$$\frac{4.28}{28.28} = 0.15$$

The answer is **(A)**

Solution T33

Regardless of certain characteristics, the roundabout capacity should be limited to 1800 vehicles per hour per lane. Since this is a single lane roundabout, the maximum is 1800 vph.

The answer is **(A)**

Solution T34

The travel speed is the length of roadway divided by the running time plus the rough vehicle delay. First calculate the delay:

$$d_t = \text{Rough vehicle delay} = t_l 60 \frac{L_{ft}}{5280} = 1.2(60)(2) = 144\ s$$

t_l = 1.2 from HCM Exhibit 18-27

Then you can calculate the travel speed:

$$S_T = \frac{3600L}{5280(t_{RT} + d_t)} = \frac{3600(2)}{(420 + 144)} = 12.8\ mph$$

The answer is **(A)**

Solution T35

The flow rate for a gutter can be determined from the following:

$$Q = K(^Z/_n)s^{1/2}y^{8/3}$$

Plugging in the variables provides:

$$Q = 0.56\left(\frac{(^1/0.025)}{0.012}\right)0.003^{1/2}0.25^{8/3} = 2.53\ cfs$$

Remember that K is taken as a constant = 0.56 ft³/(s-ft)

The answer is **(A)**

Solution T36

Let's look at each statement:

A – If the soil is weaker, a larger SN is required and therefore a larger thickness. True
B – Once the resilient modulus exceeds 30,000 psi the values are no longer applicable. False
C – CBR and R values are applicable to fine grained soils which ML, SC, and SP are. True
D – Using the appropriate equations we can correlate the numbers. M_R = 1000 + 555R = 1000 + 555(6) = 4330. M_R = 1500(CBR) = 4330: CBR = 2.9. True

The answer is **(B)**

Solution T37

The GDHS chapter 10 provides guidance on the design on interchanges. The minimum distance between interchanges is 2 miles for rural highways.

The answer is **(D)**

Solution T38

First this example requires the knowledge of the definition of the balancing point. This is where the onsite cut and fill are equal from that point to the previous or next balance point. At the start point of the project, the cut and fill are zero so this is the first balance point. We can then find the next point by seeing where the cut and fills even out. We simply need to use engineering judgement given the lack of information. The profile begins with a fill until about 130' ft and then begins the cut section. The balance point is when the cut volume equals the fill. By judgement it is toward the end of the profile. Therefore, we can take it to be about the 200' mark which is station 16+50.

The answer is **(D)**

Solution T39

The pond will overtop if the capacity is exceeded within the given time period of 6 hours. The amount that will overtop the pond is $200000 - 80000 = 120000 \ ft^3$. Then we can get an average flow rate over the 6 hr period: $120000/(6.0(60)(60)) = 5.56 \ ft^3/s$

Then we need to determine the intensity which will cause the three drainage areas to cause this flow rate. The flow into the pond is the sum of the three areas using the rational method:

$Q_T = A_1C_1i + A_2C_2i + A_3C_3i$

$5.56 = i((5)(.16) + (4)(0.23) + (6)(0.22))$; Solve for $i = 1.83$ in/hr

The answer is **(B)**

Solution T40

Present worth is determined by calculating an equivalent present value of a future cost. You can convert a future cost to a present by the following:

$$P = F(1 + i)^{-n}$$

Let's analyze each alternative

Alternate 1

Present value = 300 + $300(1 + 0.035)^{-10}$ + $300(1 + 0.035)^{-20}$ + $300(1 + 0.035)^{-30}$ = 300 + 212.7 + 150.77 + 106.9 = 770.4 Thousand dollars

Alternate 2

Present value = 650 + $300(1 + 0.035)^{-20}$ = 650 + 150.77 = 800.770 Thousand dollars

Alternative 1 is the less expensive option.

The answer is **(A)**

Answer Key

M1	A	M41	C	T1	A
M2	C	M42	D	T2	D
M3	C	M43	C	T3	A
M4	D	M44	D	T4	D
M5	A	M45	A	T5	B
M6	B	M46	B	T6	B
M7	A	M47	B	T7	C
M8	A	M48	B	T8	C
M9	C	M49	C	T9	B
M10	D	M50	C	T10	B
M11	B	M51	A	T11	C
M12	C	M52	C	T12	D
M13	D	M53	B	T13	D
M14	A	M54	D	T14	D
M15	D	M55	D	T15	B
M16	A	M56	C	T16	B
M17	C	M57	B	T17	B
M18	D	M58	D	T18	C
M19	B	M59	B	T19	C
M20	A	M60	B	T20	B
M21	C	M61	C	T21	A
M22	A	M62	A	T22	A
M23	D	M63	A	T23	D
M24	C	M64	D	T24	D
M25	A	M65	A	T25	C
M26	C	M66	C	T26	A
M27	B	M67	A	T27	D
M28	C	M68	A	T28	A
M29	A	M69	D	T29	B
M30	A	M70	B	T30	C
M31	B	M71	C	T31	D
M32	D	M72	D	T32	A
M33	D	M73	B	T33	A
M34	C	M74	C	T34	A
M35	A	M75	A	T35	A
M36	C	M76	C	T36	B
M37	C	M77	D	T37	D
M38	C	M78	D	T38	D
M39	B	M79	C	T39	B
M40	B	M80	C	T40	A

References

Civil Engineering Reference Manual for the PE Exam Fifteenth edition 2015. Michael R. Lindeburg Professional Publications Inc. (PPI)

A Policy on Geometric Design of Highways and Streets, 6th edition, 2011 (including November 2013 errata), American Association of State Highway & Transportation Officials

AASHTO Guide for Design of Pavement Structures (GDPS-4-M), 1993, and 1998 supplement, American Association of State Highway & Transportation Officials,

AASHTO Roadside Design Guide, 4th edition, 2011 (including February 2012 and July 2015 errata), American Association of State Highway & Transportation Officials,

AASHTO Mechanistic-Empirical Pavement Design Guide: A Manual of Practice, interim edition, July 2015, American Association of State Highway & Transportation Officials

AASHTO Guide for the Planning, Design, and Operation of Pedestrian Facilities, 1st edition, 2004, American Association of State Highway & Transportation Officials,

AASHTO Highway Safety Manual, 1st ed., 2010, vols. 1–3 (including September 2010, February 2012, and March 2016 errata), American Association of State Highway & Transportation Officials,

The Asphalt Handbook (MS-4), 7th edition, 2007, Asphalt Institute

Highway Capacity Manual 6th edition, vols. 1–3, Transportation Research Board—National Research Council,

Manual on Uniform Traffic Control Devices, 2009, including Revisions 1 and 2 dated May 2012, U.S. Department of Transportation—Federal Highway Administration

Design and Control of Concrete Mixtures, 16th edition, 2016, Portland Cement Association

FHWA Hydraulic Design of Highway Culverts, Hydraulic Design Series Number 5, Publication No. FHWA-HIF-12-026, 3rd edition, April 2012, U.S. Department of Transportation—Federal Highway Administration

Made in the
USA
Middletown, DE